MONOGRAPHS ON
APPLIED PROBABILITY AND STATISTICS

General Editors

M. S. BARTLETT, F.R.S., *and* D. R. COX, F.R.S.

THE STATISTICAL ANALYSIS OF SPATIAL PATTERN

The Statistical Analysis of Spatial Pattern

M. S. BARTLETT F.R.S.

Emeritus Professor of Biomathematics
University of Oxford

LONDON
CHAPMAN AND HALL

A HALSTED PRESS BOOK
JOHN WILEY & SONS INC., NEW YORK

First published 1975
by Chapman and Hall Ltd
11 New Fetter Lane, London EC4P 4EE

© *1975 M. S. Bartlett*

Typeset by Preface Ltd, Salisbury, Wiltshire
and printed in Great Britain by
Whitstable Litho, Whitstable, Kent

ISBN 0 412 14290 2

Distributed in the U.S.A. by Halsted Press,
a Division of John Wiley & Sons, Inc., New York

Library of Congress Cataloging in Publication Data

Bartlett, Maurice Stevenson,
The Statistical Analysis of Spatial Pattern

(Monographs on Applied Probability and Statistics)
Includes Bibliography and Index.
1. Spatial analysis (Statistics) I. Title.
QA278.2.B37   519.5′3   75-31673
ISBN 0 470-05467-0

Contents

Part II
Examples of statistical analyses

Preface

In a contribution (Bartlett, 1971a) to the Symposium on
Statistical Ecology at Yale in 1969, I noted in my
introductory remarks that that paper was not intended to be
in any way a review of statistical techniques for analysing
spatial patterns. My contribution to a conference at Sheffield
in 1973 aimed, at least in part, to supply such a review and
forms the basis of this monograph; but in these prefatory
remarks I must still make clear what I decided to discuss, and
what I have omitted.

Broadly speaking, the coverage is that included in seminars
and lectures I have given on this theme since 1969. We may
divide problems of spatial pattern (in contrast with complete
random chaos) into (i) detecting departures from randomness,
(ii) analysing such departures when detected, for example, in
relation to some stochastic model and (iii) special problems
which require separate consideration; for example,
sophisticated problems of pattern recognition in specific
fields, such as the computer reading of handwriting or
recognition of chromosomes.

While I shall refer to (i), I consider that the area denoted
by (ii) is now of more interest and importance, and is the one
to be mainly discussed. Class (iii) is of course of considerable
importance, but not one that is conveniently included with
(ii), nor indeed would I be the right person to attempt to
discuss problems in this class. Classes (i) and (ii) are indeed
broad enough, as is adequately indicated if we recall that the
classes of models needed in the one-dimensional case may
include all stochastic processes in time (as analogous, with
reservations to be noted, to one space dimension), and that it

is intended to concentrate mainly on the two-dimensional spatial analogues.

I have found it convenient to separate the review of theoretical models from the statistical examples, partly because, as I hope will become clear, the development of the appropriate theory is a vital part of knowing how to conduct a meaningful analysis.

While I am defining the scope of this monograph, let me mention further problems omitted for convenience, or for brevity. One is the specific problem of robust density estimation in ecology (e.g. of plants or trees) in the case of non-random pattern, even though this is linked in the literature, with the study of particular patterns and models (see, for example, Holgate, 1972, Diggle, 1973). Another is the problem of multivariate measurements of one kind or another, including the vast area of cluster and discriminant analysis techniques applied to spatial pattern problems. Finally, I shall not discuss the important problem, for example in epidemiology, usually referred to as the detection of space-time interactions.

The large area of problems remaining I shall classify in relation to the appropriate stochastic model, whether (i) continuous variable over continuous space, denoted by $X(\mathbf{r})$ (ii) a point process $N(\mathbf{r})$ or $dN(\mathbf{r})$ over continuous space, (iii) a process X_i over a lattice of sites i. We could of course divide the class of variables X_i in (iii) into continuous or discrete (e.g. binary), but part of the relevant theory at least is common, and is conveniently dealt with before any differences are noted.

Some problems, while strictly classifiable under one of these headings, may need separate consideration. An example is a mosaic pattern of vegetation over an area, which is divided into contiguous continuous regions where the vegetation is or is not present. This is classifiable as $X(\mathbf{r})$, where X is either 1 or 0. Other problems might be related to line processes rather than to point processes or to continuous processes, and some discussion of this class of processes has been included.

Finally, as in the one-dimensional case, it is hardly feasible to present standard methods for the analogues of non-stationary processes with the generality accorded to stationary models, which will be a further theoretical constraint usually in mind. This classification of models appropriate to some types of two-dimensional data is in the last resort one of expediency, and must never be assumed exhaustive.

The two-part structure has the perhaps dubious attraction that theoreticians can stop at Part I, and non-mathematical biometricians can concentrate on Part II; but to my mind the interdependence of both Parts already referred to is such that most readers, and certainly most statisticians and biometricians, should read this monograph as a whole. If they do, I hope they will find the central role played by the *same* stochastic processes and models in both physical and biological problems one of the fascinating aspects to emerge in recent years from this whole field of study (see in particular section 2.2).

March 1975   M.S.B.

Acknowledgements

It is a pleasure to record my acknowledgments to various persons and sources, including J. Gani, Editor-in-Chief of *Adv. Appl. Prob.*, for permission to make use of material in my published review ((1974a), 6, 336–58), and in particular Figs. 1, 2, 3 and 5 of that review. Acknowledgments made then to J. M. Cullen for providing biological data on nearest-neighbour distances, especially M. H. MacRoberts' data on gulls' nests, to P. D. M. Macdonald and Jennifer Brennan for assistance with its analysis, and to J. E. Besag and P. Diggle for information on their own work, are repeated here. I am especially indebted to J. E. Besag for advance information on his important work on the general theory and statistical analysis of conditional lattice systems, particularly that published in *J. R. Statist. Soc.* B 36, (1974), 192–236. Figs. 1 and 12 of this monograph are reproduced by kind permission from *Biometrika*, (1964), 51 and Fig. 7 from *J. Appl. Prob*, (1974), 11, 299–311 and 40 (1953), 287–305 originally published by the University of California Press; Fig. 2 is from *Proc. 5th Berkeley Symp. on Math. Stat. and Prob.* Vol. III (1967b), 135–53 and reprinted by permission of the Regents of the University of California. Finally, I am most grateful to Mrs K. Earnshaw (with some assistance also from Mrs Nancy Amery) for their most efficient secretarial help in the preparation of the typescript.

PART I

Survey of Underlying Theory

CHAPTER ONE

Continuous, Point and Line Processes

1.1 Continuous processes $X(\mathbf{r})$. Autocorrelation and spectral theory

For the general mathematical theory of stochastic processes, and in particular stationary processes, the reader is referred to one of the several books on stochastic processes now available (e.g. Bartlett, 1966; Cox and Miller, 1965). For a stochastic process* $\{X(t)\}$, complete stationarity implies stochastic equivalence of $\{X(t)\}$ and $\{X(t + h)\}$. It is convenient to extend the stationarity definition to processes in more than one dimension, so that even for a space vector $\mathbf{r}$ the stationarity property implies stochastic equivalence of $\{X(\mathbf{r})\}$ and $\{X(\mathbf{r} + \mathbf{h})\}$.

When the variable X has a continuous range, whether or not $\mathbf{r}$ is a continuous or discrete vector variable, autocorrelation or spectral theory provides a natural theoretical tool, in the sense that if $\{X(\mathbf{r})\}$ is a normal (Gaussian) process, the autocorrelation or equivalent spectral function is the only function that arises in the description or specification of the process, apart from a possible non-zero mean $E\{X(\mathbf{r})\} = m$. There should be no need to review this theory in complete detail, but its salient features in the case of any number of dimensions will be recalled (cf., for example, Bartlett, 1966 § § 6.5 and 9.4). Thus, taking for

---

*The entire process is strictly denoted by $\{X(t)\}$, as distinct from a single random variable $X(t)$ at time t, though in some contexts the latter notation may be used for $\{X(t)\}$ without ambiguity.

convenience the constant mean $m = E\{X(\mathbf{r})\}$ zero, we may define the autocovariance and autocorrelation functions by $w(s)$ and $\rho(s)$ respectively, where

$$\left.\begin{array}{l} w(s) = \sigma^2 \rho(s) = E\{X(\mathbf{r}+s)X^*(\mathbf{r})\}, \\ \sigma^2 = w(0), \end{array}\right\} \tag{1}$$

and $X^*(\mathbf{r})$, the complex conjugate of $X(\mathbf{r})$, is of course identical with $X(\mathbf{r})$ for real processes. The spectral distribution function $F(\omega)$ in the case, say, of two dimensions is related with $\rho(s)$ by the equation

$$\rho(s) = \int e^{i(s_1\omega_1 + s_2\omega_2)} \, dF(\omega), \tag{2}$$

where $s = (s_1, s_2)$, $\omega = (\omega_1, \omega_2)$. If $dF(\omega)$ arises entirely from a spectral density function $f(\omega)$, then (2) becomes

$$\rho(s) = \int e^{i(s_1\omega_1 + s_2\omega_2)} f(\omega) \, d\omega, \tag{3}$$

where $d\omega = d\omega_1 \, d\omega_2$. With this 'standardized' definition for $F(\omega)$ (and $f(\omega)$), note that the autocovariance function is mathematically related to $\sigma^2 F(\omega)$.

For any linear transformation or 'filter' of $X(\mathbf{r})$ of the type

$$Y(\mathbf{r}) = \int X(s) dH(\mathbf{r} - s), \tag{4}$$

it is easy to see that if

$$h(\omega) = \int e^{-i(s_1\omega_1 + s_2\omega_2)} \, dH(s), \tag{5}$$

then for $Y(\mathbf{r})$

$$\sigma_Y^2 f_Y(\omega) = h(\omega)\sigma_X^2 f_X(\omega)h^*(\omega), \tag{6}$$

where it is assumed that $f_X(\omega)$ exists. In particular, when $X(\mathbf{r})$ degenerates to a process for which $\rho(s) = 0$ for any $s \neq 0$, formula (6) becomes

$$f_Y(\omega) \propto h(\omega)h^*(\omega), \tag{7}$$

a type of formula familiar in the theory of one-dimensional time-series. A further point to note is that a strictly continuous spatial area ($X(\mathbf{r})$ might, for example, represent soil acidity varying over an area of land) can be studied from its values over a regular lattice (as in one dimension), though 'aliasing' of the higher frequencies then occurs. Thus if the two-dimensional rectangular lattice has cell dimensions unity, then the autocovariance function in (1) is only available for s changing by unit steps. Formula (2) then becomes modified to

$$\rho(s) = \int e^{i(s_1\omega_1 + s_2\omega_2)}\, dF_1(\omega), \tag{8}$$

where $F_1(\omega)$ is only defined in the range $\omega_1 = -\pi$ to π, $\omega_2 = -\pi$ to π. Since $\rho(s)$ on the left-hand-side is unchanged, we obtain the relation

$$dF_1(\omega) = \sum_{u,v=-\infty}^{\infty} dF(\omega_1 + 2u\pi, \omega_2 + 2v\pi). \tag{9}$$

Another quite common way in which a lattice process may arise from a process over continuous space is by integration over a rectangle with the lattice point as its centre, i.e. we have a derived process

$$Y(\mathbf{r}) = \int X(\mathbf{r} + \mathbf{u})d\mathbf{u} \tag{10}$$

integrated over a rectangular region centred at $\mathbf{r}$. As (10) is a special case of (4), the relation between the spectrum of $X(\mathbf{r})$ and $Y(\mathbf{r})$ is straightforward. In fact, choosing our scales to have a square area of unit dimensions, we have from formula (6)

$$\sigma_Y^2 f_Y(\omega) = \frac{\sin^2 \omega_1}{\omega_1^2} \frac{\sin^2 \omega_2}{\omega_2^2} \sigma_X^2 f_X(\omega). \tag{11}$$

Formula (11) is, however, still for continuous $\mathbf{r}$; for discrete $\mathbf{r}$, we have the further relation (9).

In the case of processes $X(\mathbf{r})$ defined only for discrete $\mathbf{r}$, notice that formulae (4) and (5) become sums, and a formula like (7) is necessarily expressible in terms of $z = e^{i\omega}$ in the case of one dimension, or $z_1 = e^{i\omega_1}$, $z_2 = e^{i\omega_2}$ in the case of two.

The consideration of possible stochastic models and their properties, while relevant to this section, is conveniently deferred until later. In particular, it will be shown from the study of lattice nearest neighbour models (see Chapter 2) that analogous Markov field models in continuous space must also be severely restricted.

1.2 Point processes $N(\mathbf{r})$. Some specific models and their distributional properties

With point processes $N(\mathbf{r})$, where it is assumed that the increment $dN(\mathbf{r})$ between $\mathbf{r}$ and $\mathbf{r} + d\mathbf{r}$ is either 0 to 1, required, for example, as models for plant or tree stands, it seems more relevant to comment first on some useful classes of models rather than on the appropriate spectral theory, bearing in mind that the non-Gaussian character of such processes necessarily limits the use of spectral properties as an adequate description.

Two broad classes of models are as follows:

A random or Poisson (two-dimensional) process, used as a basis for generating the final model in two alternative ways.

(a) (i) *Clustering models*. The individuals are taken as parents (or nuclei) of families of children (or satellites), which are usually assumed to be distributed independently, given the position $\mathbf{s}$ of the parent individual, according to some density law $f(\mathbf{r} - \mathbf{s})$, and with family size n (excluding the parent) following a distribution law $g(n)$.

(ii) *Doubly stochastic Poisson process*. Hetero-geneity or patchiness is introduced by allowing the density parameter λ of the Poisson process to be itself a continuous stationary process $\{\Lambda(\mathbf{r})\}$, say.

(b) *Contagion or inhibitory (negative contagion) models*. The basic individuals are no longer random, but are attracted or repelled by their neighbours, even to the extent of a minimum separation distance.

The distributional properties for classes (a)(i) and (a)(ii) are fairly well-known; in particular, the form of the complete characteristic functional (c.fl.) for each of these classes is known (see, for example, Bartlett, 1964b).

(i) $\quad C\{\theta(\mathbf{r})\} \equiv E\{\exp[i\int\theta(\mathbf{r})dN(\mathbf{r})]\}$

$$= \exp(\int\lambda_0[\Psi\{\theta(\mathbf{r})\} - 1]\,d\mathbf{r}), \qquad (12)$$

where $\Psi(\theta(\mathbf{r}))$ is the c.fl. stemming from one parent (parents having density λ_0). In particular, for the case of n independent offspring with probability-generation function

$$H(z) \equiv E\{e^{zN}\} = \sum_{r=1}^{\infty} g(n)z^n,$$

$$\Psi\{\theta(\mathbf{r})\} = H(1 + \int f(\mathbf{u} - \mathbf{r})[z(\mathbf{u}) - 1]\,d\mathbf{u}), \qquad (13)$$

if $z(\mathbf{r}) = \exp\{i\theta(\mathbf{r})\}$ and only offspring are considered (parents being omitted) in $C\{\theta(\mathbf{r})\}$.

(ii) $\quad C\{\theta(\mathbf{r})\} = E_\Lambda\{\exp(\int\Lambda(\mathbf{u})[z(\mathbf{u}) - 1]\,d\mathbf{u})\}, \qquad (14)$

and in particular if

$$\Lambda(\mathbf{r}) = \int\beta(\mathbf{r} - \mathbf{v})dM(\mathbf{v}), \qquad (15)$$

where $M(\mathbf{r})$ is another basic Poisson process with density λ_1, then

$$\log C\{\theta(\mathbf{r})\} = \int\lambda_1(\exp\{\int\beta(\mathbf{r} - \mathbf{v})[z(\mathbf{r}) - 1]\,d\mathbf{r}\} - 1)d\mathbf{v}. \qquad (16)$$

Such formulae determine in principle all distributional properties of $N(\mathbf{r})$. Thus it has been noted by Paloheimo (see discussion to Matérn, 1971) that they enable nearest-neighbour distributions to be determined in the case of these non-random models. However, it should perhaps be noted also, for those unfamiliar with the use of characteristic functionals, that direct probability derivations of such distributions are not difficult to write down.

Let us take the case of the clustering model (a) (i). It is well-known that, while the sampling distribution of nearest-neighbour distance is in the purely random case the same whether this distance is from I. a random point II. a random individual, this is no longer so for a non-random process. So consider cases I and II separately, leaving aside the problem of how to pick a random individual in case II!

Case I   In this case we shall consider the distribution both if parents are excluded, and if they are included. In the former situation, consider an arbitrary region of area A, and let the probability of no children in A be P_1, and the probability of no individuals $P_2 \leq P_1$. Suppose a parent is at s, s + ds relative to the random point at the origin, and denote the event

'No children in A from parent in ds'

by $\&(ds)$. (Given the shape and position of the region A, the probability of this event can always be written down as a mathematical integral.) Then

$$P_1 = \prod_{ds} [P\{\&(ds)\}P\{\text{parent at ds}\}$$
$$+ 1 - P\{\text{parent at ds}\}] \qquad (17)$$
$$= \exp(\int \lambda_0 \, ds[P\{\&(ds)\} - 1]);$$

$$P_2 = P\{\text{no parents in } A\}P\{\text{no children in } A \mid$$
$$\text{no parents in } A\} \qquad (18)$$

$$\log P_2 = -\lambda_0 A + \int \lambda_0 \, ds[P\{\&(ds)\} - 1],$$

where the integration for $\log P_2$ is over the entire space excluding A. Note that

$$P_2 \neq e^{-\lambda A}P_1.$$

To obtain the cumulative distribution function $F(R)$ of the nearest-neighbour distance R, say, we choose a circle, radius R, centre O, and then

$$P_1 \text{ (or } P_2) = 1 - F(R). \qquad (19)$$

In Case II, we consider for simplicity the case of children only in $N(\mathbf{r})$. Let the origin O now be at a 'random individual'; any individual in A may now belong to the same family as the individual at O or to a different family. *Given* the second contingency, the relevant probability of no individuals in A is P_1 above. Let the overall probability be P_3. Then

$P_3 = P\{\text{no children of different family}\}$

$\quad \times P\{\text{no children of same family} \mid \text{no children of}$

$\quad\quad\quad\quad\quad\quad\quad\quad\quad\quad\quad \text{different family}\}$

$\quad = P\{\text{no children of different family}\}P\{\text{no children of}$

$\quad\quad\quad\quad\quad\quad\quad\quad\quad\quad\quad\quad \text{same family}\}$

as these probabilities are independent

$$= P_1 P_4 \ (\leq P_1). \tag{20}$$

where

$P_4 = P\{\text{no children of same family in } A \mid \text{child in } \mathrm{d}a \text{ at } 0\}$

$$= \frac{P\{\text{no children of same family in } A, \text{child in } \mathrm{d}a\}}{P\{\text{child in } \mathrm{d}a\}}$$

$$= \frac{P\{\text{no children in } A'\} - P\{\text{no children in } A\}.}{P\{\text{child in } \mathrm{d}a\}}$$

$$= \frac{\int [P\{\mathcal{E}'(\mathrm{d}s)\} - P\{\mathcal{E}(\mathrm{d}s)\}]\lambda_0 \,\mathrm{d}s}{\lambda \,\mathrm{d}a}, \tag{21}$$

where $\mathcal{E}'(\mathrm{d}s)$ is the event 'no children in $A' \equiv A$ excluding $\mathrm{d}a$ at 0, from parent in $\mathrm{d}s$', and $\lambda = \lambda_0 E\{N\}$.

1.2.1 *Spectral theory*

The spectral theory for two-dimensional point processes, which was given by Bartlett (1964b), is recapitulated here. Let

$$\left.\begin{array}{l} E\{\mathrm{d}N(\mathbf{r})\} = \lambda \mathrm{d}\mathbf{r}, \\ E\{\mathrm{d}N(\mathbf{r}) \,\mathrm{d}N(\mathbf{r}')\} = \{\lambda^2 + w(\mathbf{r} - \mathbf{r}')\}\mathrm{d}\mathbf{r}\mathrm{d}\mathbf{r}', (\mathbf{r} \neq \mathbf{r}'), \end{array}\right\} \tag{22}$$

where $\mathrm{d}\mathbf{r} \equiv \mathrm{d}x\mathrm{d}y$, and $\mathrm{d}N(\mathbf{r}) = N(\mathbf{r} + \mathrm{d}\mathbf{r}) - N(\mathbf{r})$. It is assumed that $\mathrm{d}N(\mathbf{r})$ is restricted to values 1 or 0 so that

$$E\{[\mathrm{d}N(\mathbf{r})]^2\} = E\{\mathrm{d}N(\mathbf{r})\}.$$

The complete covariance density function for $N(\mathbf{r})$ is then defined by $w(\mathbf{r} - \mathbf{r}') + \delta(\mathbf{r} - \mathbf{r}')$, where $\delta(\mathbf{u})$ is the two-dimensional Dirac delta-function. The corresponding complete spectral density function is defined by

$$f(\boldsymbol{\omega}) = \frac{1}{(2\pi)^2}\ [\textstyle\int e^{-i\omega'\mathbf{r}}\ w(\mathbf{r})d\mathbf{r} + \lambda] \tag{23}$$

where $\omega'\mathbf{r} = \omega_1 x + \omega_2 y$. For convenience, $4\pi^2 f(\boldsymbol{\omega})$ may be denoted by $g(\boldsymbol{\omega})$; being a 'modulation' of the uniform function λ, it cannot, as for a continuous process $X(\mathbf{r})$, be standardized to integrate to unity.

To determine $g(\boldsymbol{\omega})$ for a clustering model as in (i), let the distribution of number of offspring per parent be p_s, and let the spatial distribution function of each offspring about its parent have a density function $h(\mathbf{r})$. Then the contributions to the proper component of the covariance density are

$$\lambda_0\ \{[h(\mathbf{r}) + h(-\mathbf{r})]E\{s\} + h_2(\mathbf{r})E\{s(s-1)\}\} \tag{24}$$

if parents are included, and the last term only if parents are excluded, where $h_2(\mathbf{r})$ represents the distribution of vector distance between two offspring from the same parent. In the case of offspring only, this leads to the spectral function

$$g(\boldsymbol{\omega}) = \lambda\ \{1 + C(\boldsymbol{\omega})C^*(\boldsymbol{\omega})E\{s(s-1)\}/E\{s\}\}, \tag{25}$$

where $C(\boldsymbol{\omega})$ is the characteristic function of $h(\mathbf{r})$. In particular, for the isotropic Gaussian density

$$h(\mathbf{r}) = \frac{1}{2\pi\sigma^2}\ e^{-\frac{1}{2}\mathbf{r}^2/\sigma^2}, \tag{26}$$

so that

$$C(\boldsymbol{\omega}) = e^{-\frac{1}{2}\sigma^2(\omega_1^2 + \omega_2^2)}, \tag{27}$$

we obtain

$$g(\boldsymbol{\omega}) = \lambda\left\{1 + \frac{(m_0^2 + v_0 - m_0)e^{-\sigma^2(\omega_1^2 + \omega_2^2)}}{m_0}\right\}, \tag{28}$$

where m_0 and v_0 are mean and variance of s. For example, if

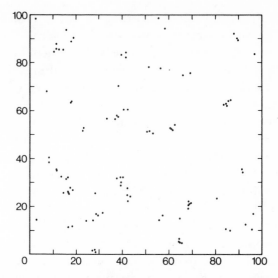

Fig. 1 *Simulation of clustering model (for detailed specification, see text). (From Biometrika, 1964b)*

p_s is Poisson, $v_0 = m_0$ and this becomes

$$g(\omega) = \lambda \{ 1 + m_0 e^{-\sigma^2 (\omega_1^2 + \omega_2^2)} \} . \tag{29}$$

A simulated example of 100 points of this last clustering process, with $m_0 = 2$ and $\sigma = 1$, is shown in Fig. 1. It should be noted that for these models the additional spectral component from $w(\mathbf{r})$ is essentially positive, as it is for doubly-stochastic Poisson models, so that it may be difficult (or even in certain cases impossible) to discriminate between these two types of model. By contrast, models involving inhibition between individuals (see § 1.2.2) can lead to negative components, as I have demonstrated elsewhere in the one-dimensional spectral analysis of traffic patterns.

1.2.2 *Contagion and inhibitory models*

Coming now to the second broad class of models referred to in § 1.2.1 viz. contagion processes, we may define these as models where individuals attract or repel each other, just as

physical particles may do. Unfortunately while these models
are very relevant for many physical and biological situations,
they are extremely difficult to cope with mathematically,
especially in more than one dimension, so that models may
have to be simulated before data can be compared with the
theoretical model.

Even in one dimension the distinction between two-way
spatial symmetry and the one-way temporal direction may be
crucial. As an example consider that of swallows perching on
a telegraph wire, and suppose that in our model we wish to
have an inhibitory minimum distance a, but otherwise no
restriction. Two alternative models would then be as follows:

(i) Apart from the minimum distance a, the birds alight at
random unidirectionally, the overall space density being λ.
In other words, apart from the 'dead intervals' a, the positions
of the birds would define a Poisson process. If the density of
such a process is λ_0, then obviously

$a + 1/\lambda_0 = 1/\lambda.$

(ii) The birds alight consecutively at random, but within a
given distance L, and still subject to the minimum distance a.
The distributional problem is now related with a classical
packing problem (the 'car-parking' problem: see, for example,
Moran, 1966; the distinction between (i) and (ii) in the
present context has been stressed and further discussed by
Diggle, 1973).

1.3 Line processes

To a considerable extent the specification of stochastic
processes is a matter of convenience. Renewal processes, for
example, are familiar in their own right, or as a special class
of point processes. If vehicles are moving on a road, they may
be represented (approximately) either as point processes
evolving in time, or as paths in a space-time continuum. The
latter static representation is an example of a line process
(Bartlett, 1967b), and clearly all examples of random paths
in space could be so designated. I shall, however, bearing in

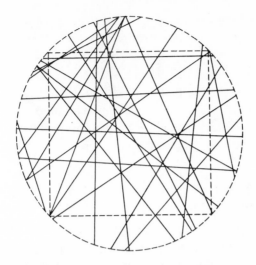

Fig, 2 *Simulation of random line process. (From Proc. 5th Berkeley Symp. on Math. Stat. and Prob., Vol. 111, 1967b)*

mind the scope of this book, restrict the class of line processes considered very drastically to straight lines in a plane. With this restriction, the number of possible examples is rather sparse, and only an introductory discussion of types of analysis will be attempted. Even the example of vehicles on a road will be excluded as being largely one-dimensional if the data consist of a single sample of vehicles passing a spatial point, whether or not velocities at this point are recorded (for analyses of such traffic data, see Bartlett, 1966 §9.23; 1967b). The existence of alternative representations of the same process we shall see becomes very relevant—for example, with regard to the appropriate spectral theory.

   Fig. 2 shows a simulated sample of random lines in a plane (for its construction, see §3.3). In order to consider a rather wider class, define first the spatial process

$$X(\mathbf{r}) = \int \xi(\mathbf{s} - \mathbf{r}) \, dN(\mathbf{s}), \tag{30}$$

where $\{N(\mathbf{s})\}$ is some point process, and $\{\xi(\mathbf{r})\}$ is a random function associated with each point event of $\{N(\mathbf{s})\}$. The $\{\xi(\mathbf{r})\}$ are independently realized for each such point. For our restricted class of line processes, $\xi(\mathbf{r})$ will be zero except on a straight line, possibly infinite in extent, where it is given a constant measure.

For such processes the direct spectral function of $X(\mathbf{r})$ may be obtained by the use of formulae like (6) and (7) of §1.1, with appropriate scaling to give a non-zero and finite measure. Thus for $N(\mathbf{s})$ in (30) a random Poisson process with density λ, and randomly orientated lines,

$$
\begin{aligned}
&\sigma_X^2 f_X(\omega) \to 2\lambda l \int_0^l (1 - z/l) J_0(\omega z) \mathrm{d}z, \\
&(\omega^2 = \omega_1^2 + \omega_2^2)
\end{aligned}
\tag{31}
$$

where l is the (constant) length of the lines, and the constant measure assigned to an element ds of a line is ds/α, where α is the 'width' of the line and tends to zero. In the further case of $l \to \infty$, this formula becomes

$$
\sigma_X^2 f_X(\omega) \to 2\lambda/\omega, \qquad (\omega > 0)
\tag{32}
$$

the measure being now $ds/\alpha\sqrt{l}$ (cf. Bartlett, 1966, §6.52).

However, these direct spectral functions do not seem especially convenient, at least for digital computations on these processes, and transformation to pure point process representations before analysis seems likely to be more convenient. Thus in the case of (effectively) infinite straight lines, these may be defined by the equations

$$
x \cos \theta + y \sin \theta = p,
$$

and specified by a point process in the infinite strip p from $-\infty$ to $+\infty$, and θ from 0 to π (cf. Kendall and Moran, 1963). If the lines have direction, then θ would vary from 0 to 2π.

If the lines are finite but constant in length, the representation (1) may be used to specify the point process $N(\mathbf{s})$ for their centres, and θ from 0 to π for their slope.

Notice that the point process representation, while two-dimensional, has a different structure in that one coordinate is an angle variable with a Fourier *series* spectrum. The combined spectrum for p (or s in the finite length case) and θ will consequently be a series of coefficients, each coefficient being a spectral function for p (or s). For completely random lines, these spectral values will be theoretically constant.

One-dimensional transect sampling In some cases it may be of interest to consider the properties of the point process obtained by intersection with a standard transect line, specified by, say p_0, θ_0 (cf. Pielou's example, §3.1). For lines of infinite extent, some useful results have been stated by Solomon and Wang (1972).

Suppose the line process is such that p is random (density μ), but θ independently distributed with cumulative distribution function $G(\theta)$. Then the point intersections on the line (p_0, θ_0) form a Poisson process with constant density $\lambda(\theta_0)$, where

$$\lambda(\theta_0) = \mu \cos \theta_0 \int_{-\infty}^{\infty} | v - \tan \theta_0 | (1 + v^2)^{-\frac{1}{2}} dH(v),$$

$$(v = \tan \theta). \quad (33)$$

In the particular case of θ random,

$$dH(v) = \frac{1}{\pi} \frac{dv}{1 + v^2}. \quad (34)$$

Special cases of (p_0, θ_0) are relevant to applications to traffic motion, where the second original dimension is time, and if v is velocity θ_0 corresponding to a sample in space is $\frac{1}{2}\pi$ and θ_0 corresponding to a sample in time is 0. Formula (33) gives in the two cases

$$\lambda(\tfrac{1}{2}\pi) = \mu \int_{-\infty}^{\infty} (1 + v^2)^{-\frac{1}{2}} dH(v), \quad (35)$$

$$\lambda(0) = \mu \int_{-\infty}^{\infty} | v | (1 + v^2)^{-\frac{1}{2}} dH(v). \quad (36)$$

For θ random, formula (33), and its particular cases (35) and (36), all give the *same* density

$$\lambda(\theta_0) = 2\mu/\pi, \quad (37)$$

whatever θ_0. This is not of course true for arbitrary $G(\theta)$, a comment which is relevant if these line-process models are being studied by sample transect lines (again cf. Pielou's example, §3.1).

Spectral functions for non-random line processes   The effect of a non-random point process e.g. a clustering process for the centres of lines of finite length, may also be introduced. Thus for lines of constant length l orientated at random the *additional* term to formula (31) is

$$4\left[\int_0^{\frac{1}{2}l} J_0(\omega u)du\right]^2 [g_N(\omega) - \lambda], \qquad (38)$$

where $g_N(\omega)$ is $4\pi^2 f(\omega)$,, where $f(\omega)$ was defined for a point $dN(s)$ by equation (23) of §1.2.1 and illustrated by formulae (28) and (29) of §1.2.1 in the case of clustering point processes.

Consider now the spectral function for s, θ (not p, θ, as l is finite). This is clearly uniform for θ, and, for s, simply $g(\omega)$. For infinite l and the p, θ representation, however, we encounter some difficulties. Suppose we consider a sample circle C_1 of radius a and the generation of lines intersecting it from a much larger ring of radius A, $A + \delta A$. For a density λ of foci generating lines, and the lines orientated at random, the number intersecting the sample circle will be $2\pi\lambda A(\delta A)$ $(\alpha/2\pi)$, where $\sin \frac{1}{2}\alpha = a/A$, $\backsim 2\lambda a(\delta A)$, and tends to infinity as we add contributions over increasing A. Thus all the lines

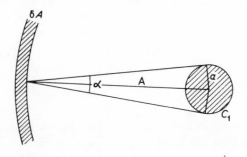

Fig. 3

intersecting the sample circle, if we reduce the density to an extent for this number to be finite, will be coming from foci infinitely distant. This implies a breakdown of the model, and the construction of more realistic models would be required when the conditions in any particular problem were specified in detail.

CHAPTER TWO

Nearest-neighbour Systems on a Lattice

2.1 Lattice processes X_i. Continuous variables. Conditional and simultaneous systems

For definiteness, we shall usually consider uniformly-spaced rectangular lattice sites $i = (r, s)$ in two dimensions, and in this section we suppose that the associated random variable X_i has a continuous range of possible values. The spectral and autocorrelation theory recalled in §1.1 is then modified as indicated in equation (8) of that section, viz

$$\rho(s) = \int e^{i(s_1 \omega_1 + s_2 \omega_2)} \, dF_1(\omega) \tag{1}$$

over the range $-\pi$ to π for ω_1 and ω_2. As before, we denote the autocovariance function by $w(s) = \sigma_X^2 \rho(s)$. For a linear transformation

$$Y_{rs} = \Sigma_{u,v} X_{u,v} h_{r-u, s-v} \tag{2}$$

equations (6) and (7) of §1.1 are not altered provided now

$$h(\omega) = \Sigma_{u,v} e^{-i(u\omega_1 + v\omega_2)} h_{r-u, s-v}. \tag{3}$$

By analogy with auto-regressive and moving average models in time series, an important class of models will be those where dependence is local (e.g. nearest-neighbour). The most familiar will be linear, but are of *two* possible forms (cf. Brook, 1964). The first of these is the *simultaneous* model

$$X_{rs} = \beta_1 X_{r-1,s} + \beta_1' X_{r+1,s} + \beta_1' X_{r,s-1} + \beta_2' X_{r,s+1} + Y_{rs} \tag{4}$$

where $\{Y_{rs}\}$ is an array of independent variables with identical distribution (and $E\{Y\} = 0$). In particular, in the laterally symmetric case, this model becomes

$$X_{rs} = \beta_1 (X_{r-1,s} + X_{r+1,s}) + \beta_2 (X_{r,s-1} + X_{r,s+1}) + Y_{rs}$$

$$(5)$$

Whether or not the Y_{rs} are normal, the spectral function is easily written down, owing to the uniform spectrum of Y_{rs} and the linear relation (5), say. Thus the spectral function for a linear relation

$$\phi(E_1, E_2) X_{rs} = Y_{rs}, \tag{6}$$

where $E_1 = 1 + \Delta_1$ is the shift difference operator acting on r and E_2 similarly for s, since formally

$$X_{rs} = \phi^{-1}(E_1, E_2) Y_{rs},$$

and E_1 is equivalent to $e^{i\omega_1}$ for the Fourier transform (3), and and E_2 to $e^{i\omega_2}$, must be proportional to

$$[1 - 2(\beta_1 \cos \omega_1 + \beta_2 \cos \omega_2)]^{-2} \tag{7}$$

(cf. similar one-dimensional results in Bartlett (1966), §6.3).

One difficulty with the model represented by equation (5) is that, even if the Y_{rs} and hence the X_{rs}, are normal, the estimation of β_1 and β_2 from data is not a simple least-squares problem, owing to the complicated Jacobian when transforming from the independent Y_{rs} to the observed X_{rs} (cf. Whittle (1954); Ord (1972), and examples in §4.1).

In order to obtain an asymptotic expression for this Jacobian, we shall follow the derivation due to Whittle. We assume that $\log f(\omega)$ has a Fourier expansion

$$\log G(z_1, z_2) = - \left(\tfrac{1}{2}\alpha_{00} + \sum_{k=1}^{\infty} \alpha_{0k} z_2^k + \sum_{j=1}^{\infty} \sum_{k=-\infty}^{\infty} \alpha_{jk} z_1^j z_2^k \right),$$

$$(8)$$

so that

$$f(\omega) \propto 1/[G(\omega)G^*(\omega)], \tag{9}$$

which implies that the autoregressive scheme

$$\phi(\mathbf{E})X_{rs} = \epsilon_{rs}, \qquad (\sigma_\epsilon^2 = v), \tag{10}$$

where $\mathbf{E} \equiv (E_1, E_2)$, and $\phi(\mathbf{E})$ is the one-sided spatial operator equivalent to $G(\omega)$, will have the required spectral function $f(\omega)$. We assume here that $G(\omega)$ has a Fourier expansion.

Under the assumption of ϵ_{rs} normal, their joint probability is

$$(2\pi v)^{-\frac{1}{2}mn} \exp\{-\tfrac{1}{2}\Sigma_{r,s}\epsilon_{rs}^2/v\}\Pi_{r,s}\mathrm{d}\epsilon_{rs}$$

for a rectangular array of mn observations. Hence, neglecting end effects, we have for X_{rs} the probability

$$\alpha^{mn}(2\pi v)^{-\frac{1}{2}mn} \exp\{-\tfrac{1}{2}\Sigma_{r,s}\phi(\mathbf{E})X_{rs}/v\}\Pi_{r,s}\mathrm{d}X_{rs} \tag{11}$$

where α is the coefficient of X_{rs} in $\phi(\mathbf{E})X_{rs}$, and is

$$\exp(-\tfrac{1}{2}\alpha_{0\,0}) = \exp\left\{-\frac{1}{8\pi^2}\int \log f(\omega)\mathrm{d}\omega\right\}.$$

The logarithm of the likelihood function may thus be written

$$\tfrac{1}{2}mn(2\log\alpha - \log v) - \tfrac{1}{2}\Sigma_{r,s}\epsilon_{rs}^2/v), \tag{12}$$

or, after maximization with respect to v,

$$\tfrac{1}{2}mn(2\log\alpha - \log\hat{v}) - \tfrac{1}{2}mn,$$

where $\hat{v} = \Sigma_{r,s}\epsilon_{rs}^2/(mn)$. Hence we require to minimize $k\hat{v}$, where $k = 1/\alpha^2$, i.e.

$$\log k = \frac{1}{4\pi^2}\int \log f(\omega)\mathrm{d}\omega. \tag{13}$$

Conditional systems　The second model, while still to be linear *for continuous variables* X_i, is defined as a *conditional* nearest-neighbour model

$$P\{x_{rs} \mid \text{all other values}\}$$

$$= P\{x_{rs} \mid x_{r-1,s}, x_{r+1,s}, x_{r,s-1}, x_{r,s+1}\}, \tag{14}$$

and is also known as a *Markov field*, though it should be

```
 .    .    3    .    .
 .    3    2    3    .
 3    2    1    2    3
 .    3    2    3    .
 .    .    3    .    .
```

Fig. 4

remarked that the close relation between ordered Markov sequences and conditional nearest-neighbour systems in one dimension (Bartlett, (1966) §2.22) is less useful in more than one dimension, as the ordering is more complicated and not unique. It may sometimes be convenient, for example, to consider the particular ordering shown in the diagram (Fig. 4), starting with the first set $S_1 \equiv x_{rs}$ at position 1, the second set $S_2 \equiv (x_{r-1,s}, x_{r+1,s}, x_{r,s-1}, x_{r,s+1})$ at positions 2, and so on, but notice that only the first set in this hierarchy consists of a single variable.

The requirement (4), while a natural possible assumption, turns out to be a severe one. It may be shown that if we define

$$r(x \mid y,u,v,w) = \log \left\{ \frac{P(x \mid y,u,v,w)}{P(0 \mid y,u,v,w)} \right\} \qquad (15)$$

where without loss of generality 0 is assumed a feasible value of x, and $x,(y,u,v,w)$ are realized values of S_1, S_2, then (as a special case of a more general result derived in §2.3) $r(x \mid y,u,v,w)$ must have the form

$$x\{\phi(x) + y\psi_1(x,y) + u\psi_1(u,x) + v\psi_2(x,v) + w\psi_2(w,x)\}; \qquad (16)$$

and with lateral symmetry, so that $r(x \mid y,u,v,w)$ is a function of $y + u$ and $v + w$, then $r(x \mid y,u,v,w)$ becomes of the form

$$x\{\phi(x;\alpha_1, \alpha_2) + \alpha_1(y + u) + \alpha_2(v + w)\}. \qquad (17)$$

From this result Besag (1972b) has shown that the conditional cumulant function $K(\theta)$ for X must be quadratic in θ, and X normal, provided the model is linear i.e.

$$E\{X \mid y,u,v,w\} = \gamma + \beta_1(y + u) + \beta_2(u + w). \qquad (18)$$

For if the *conditional* moment-generating function of X, given y, u, v, w, is $M(\theta) \equiv E\{e^{X\theta}\}$, and we denote $y + u$ by λ, $v + w$ by μ, and $\exp r(x)$ summed over all x by $C(\lambda, \mu)$, we may note that

$$K(\alpha_1 \theta) \equiv \log M(\alpha_1 \theta) = \log C(\lambda + \theta, \mu) - \log C(\lambda, \mu),$$

and as equation (18) linear in λ and μ must result from $\partial \log C/\partial(\alpha_1 \lambda)$, the quadratic form in θ for $K(\theta)$ readily follows.

A more general formulation of conditional nearest-neighbour lattice schemes is deferred until the end of this chapter (§2.3).

Linear conditional spatial-temporal models   Consider next the class of linear conditional spatial-temporal models (cf. Bartlett, 1971b)

$$dX_{rst} = - \lambda\phi(\mathbf{E})X_{rst}\, dt + dZ_{rst}, \tag{19}$$

where $\phi(\mathbf{E})$ is for the moment any linear spatial displacement operator acting on X_{rst}, and dZ_{rst} are homogeneous independent terms (with zero means) for all r, s and t, $t + dt, \ldots$. Any such process, if it leads to eventual stationarity (in both space and time), leads to the following consequences:

(i) the spatial-temporal spectrum $f_X(\omega_t, \boldsymbol{\omega})$ must be proportional to

$$[i\omega_t - \lambda G(\boldsymbol{\omega})]^{-1} [- i\omega_t - \lambda G^*(\boldsymbol{\omega})]^{-1}, \tag{20}$$

where $G(\boldsymbol{\omega})$ is the Fourier factor corresponding to the linear operator ϕ, and $f_X(\omega_t, \boldsymbol{\omega})$ is defined similarly to $f_X(\boldsymbol{\omega})$ (for two lattice space dimensions) by

$$\sigma_X^2 f_X(\omega_t, \boldsymbol{\omega}) = \frac{1}{(2\pi)^3} \int_{-\infty}^{\infty} \sum_{-\infty} \sum_{-\infty} e^{-i(\omega_t \tau + \omega_1 u + \omega_2 v)}$$

$$\times w(\tau, u, v)d\tau \tag{21}$$

where

$$w(\tau, u, v) = E\{X_{rst}X_{r+u,s+v,t+\tau}\}; \tag{22}$$

(ii) the marginal spatial spectrum $f_X(\omega)$ is consequently proportional to

$$\{\tfrac{1}{2}\lambda[G(\omega) + G^*(\omega)]\}^{-1}; \tag{23}$$

(iii) if

$$(\phi(\mathbf{E}) - 1)X_{rs} = \beta_1' X_{r-1,s} + \beta_1'' X_{r+1,s} + \beta_2' X_{r,s-1}$$
$$+ \beta_2'' X_{r,s+1} \tag{24}$$

then from (ii) we note that the marginal spectrum for the linear case (24) is proportional to

$$[1 - 2\beta_1 \cos \omega_1 - 2\beta_2 \cos \omega_2]^{-1}, \tag{25}$$

where $\beta_1 = \tfrac{1}{2}(\beta_1' + \beta_1'')$, $\beta_2 = \tfrac{1}{2}(\beta_2' + \beta_2'')$. This is in contrast with result (7), which has negative index *two*. Note further that the above choice of $\phi(\mathbf{E}) - 1$ to correspond to the nearest neighbours of X_{rs} does not in itself ensure that the marginal spatial distribution is a conditional nearest-neighbour system as earlier defined. In fact, for binary X_{rs} it is known that it is not (see §2.2).

For normal X_{rs} (and $\beta_1' = \beta_1''$, $\beta_2' = \beta_2''$), however, Besag's results imply that it is, and it is termed the *auto-normal* model. In this case the result in equation (18), which for $E\{X\} = 0$ becomes

$$E\{X_{rs} \mid y,u,v,w\} = \beta_1(y + u) + \beta_2(v + w) \tag{26}$$

is consistent with the spectrum (25).

One-dimensional simultaneous and conditional systems   The contrast between the processes (5) and (26), demonstrated by their different spectral functions, has not always been appreciated, and it may therefore be helpful to demonstrate it even in the one-dimensional case. Let the analogue of (5) in one dimension be

$$X_r = \beta_1(X_{r-1,s} + X_{r+1,s}) + Y_r, \tag{27}$$

with spectral function proportional to

$$[1 - 2\beta_1 \cos \omega_1]^{-2}. \tag{28}$$

The equivalence already noted in one dimension between a Markov sequence X_r and a conditional nearest-neighbour system leads for the linear Markov process

$$X_r = \rho X_{r-1} + Y_r \tag{29}$$

to the equivalent regression relation

$$X_r = \frac{\rho}{1 + \rho^2} \, [X_{r-1} + X_{r+1}] + Z_r, \tag{30}$$

but this equation, which corresponds to (26), must *not* be identified with (27) (with $\beta_1 = \rho/(1 + \rho^2)$). The spectrum of (30) must of course be identical with that of (29) viz

$$\frac{\sigma_Y^2}{\sigma_X^2} \frac{1}{(1 - e^{i\omega})(1 - \rho e^{-i\omega})} = \frac{\sigma_Y^2}{\sigma_X^2(1 + \rho^2)[1 - 2\beta_1 \cos \omega]}. \tag{31}$$

This may be reconciled with the form of equation (30) by noting that the Z_r in (30) are not mutually independent (like the Y_r in (27)), but are defined in terms of the independent Y_r in (29) by the relation

$$Z_r = (Y_r - \rho Y_{r+1})/(1 + \rho^2). \tag{32}$$

It can thus be misleading to refer to such a model as (27) as a regression model, and it might be better to term it a moving-average model; it is, however, equivalent to the one-sided second-order autoregressive model

$$X_r = 2\alpha X_{r-1} - \alpha^2 X_{r-2} + W_r, \text{ where } \beta_1 = \alpha/(1 + \alpha^2). \tag{33}$$

Conversion of simultaneous system to conditional system   It is also clear from the more general derivation of normal conditional schemes for linear dependence extending beyond first neighbours that simultaneous normal systems may be converted to equivalent conditional systems. For example, the scheme (27) (or (33)) has spectrum

$$[1 - \beta_1(z_1 + z_1^{-1})]^{-2}, \quad (z_1 \equiv e^{i\omega_1}),$$

which on expansion yields

$$[1 + 2\beta_1^2 - 2\beta_1(z_1 + z_1^{-1}) + \beta_1^2(z_1^2 + z_1^{-2})]^{-1} \qquad (34)$$

which must correspond to the conditional scheme

$$E\{X_r \mid \text{all other values}\} = \gamma_1(X_{r-1} + X_{r+1})$$
$$+ \gamma_2(X_{r-2} + X_{r+2}) \qquad (35)$$

where

$$\gamma_1 = 2\beta_1/(1 + 2\beta_1^2), \gamma_2 = -\beta_1^2/(1 + 2\beta_1^2).$$

Similarly in two dimensions the simultaneous scheme with spectrum

$$[1 - \beta_1(z_1 + z_1^{-1}) - \beta_2(z_2 + z_2^{-1})]^{-2}$$
$$= [1 + 2\beta_1^2 + 2\beta_2^2 - 2\beta_1(z_1 + z_1^{-1}) - 2\beta_2(z_2 + z_2^{-1})$$
$$+ \beta_1^2(z_1^2 + z_1^{-2}) + \beta_2^2(z_2^2 + z_2^{-2})$$
$$+ \beta_1\beta_2(z_1 z_2 + z_1^{-1} z_2 + z_1 z_2^{-1} + z_1^{-1} z_2^{-1})]^{-1}, \qquad (36)$$

corresponds to the conditional scheme

$$E\{X_{rs} \mid \text{all other values}\} = \gamma_1(X_{r-1,s} + X_{r+1,s})$$
$$+ \gamma_2(X_{r,s-1} + X_{r,s+1}) + \gamma_3(X_{r-2,s} + X_{r+2,s})$$
$$+ \gamma_4(X_{r,s-2} + X_{r,s+2})$$
$$+ \gamma_5(X_{r-1,s-1} + X_{r-1,s+1} + X_{r+1,s-1} + X_{r+1,s+1})$$
$$\qquad (37)$$

where

$$\gamma_1 = \frac{2\beta_1}{1 + 2\beta_1^2 + 2\beta_2^2}, \gamma_2 = \frac{2\beta_2}{1 + 2\beta_1^2 + 2\beta_2^2},$$

$$\gamma_3 = \frac{-\beta_1^2}{1 + 2\beta_1^2 + 2\beta_2^2}, \gamma_4 = \frac{-\beta_2^2}{1 + 2\beta_1^2 + 2\beta_2^2},$$

$$\gamma_5 = \frac{\beta_1\beta_2}{1 + 2\beta_1^2 + 2\beta_2^2}$$

Approximate conversion of two-sided conditional system to one-sided system The equivalence of the one-dimensional two-sided system (30) to the one-sided system (29) is important for statistical inference purposes (see Part II), as it implies that the simultaneous probability of the observed values (i.e. the likelihood function when regarded as a function of the parameters of the system) readily factorizes. This is no longer possible in more than one dimension (cf., however, the use of coding methods below), but one-sided approximations may be useful if the regression parameters are not too large. Thus for the scheme with spectrum

$$[1 - \beta_1(z_1 + z_1^{-1}) - \beta_2(z_2 + z_2^{-1})]^{-1} \tag{38}$$

we have the successive approximations:

(i) $\quad X_{rs} = b_1 X_{r-1,s} + b_2 X_{r,s-1} + Y_{rs}, \tag{39}$

with exact spectrum for comparison with (38)

$$
\begin{aligned}
(1 - b_1 z_1 - b_2 z_2)^{-1}&(1 - b_1 z_1^{-1} - b_2 z_2^{-1})^{-1} \\
= [1 + b_1^2 + b_2^2 &- b_1(z_1 + z_1^{-1}) - b_2(z_2 + z_2^{-1}) \\
&+ b_1 b_2(z_1 z_2^{-1} + z_1^{-1} z_2)]^{-1}.
\end{aligned} \tag{40}
$$

(ii) $\quad X_{rs} = b_1 X_{r-1,s} + b_2 X_{r,s-1} + b_1 b_2 X_{r+1,s-1} + Y_{rs}, \tag{41}$

with exact spectrum

$$
\begin{aligned}
(1 - b_1 z_1 &- b_2 z_2 - b_1 b_2 z_2 z_1^{-1})^{-1} \\
(1 - b_1 z_1^{-1} &- b_2 z_2^{-1} - b_1 b_2 z_2^{-1} z_1)^{-1} \\
= [1 + b_1^2 &+ b_2^2 + b_1^2 b_2^2 - b_1(a - b_2^2)(z_1 + z_1^{-1}) \\
&- b_2(z_2 + z_2^{-1}) + b_1^2 b_2(z_1^2 z_2^{-1} + z_1^{-2} z_2)]^{-1}.
\end{aligned} \tag{42}
$$

Thus the identification in the case of (i) of β_1 with $b_1/(1 + b_1^2 + b_2^2)$ and β_2 with $b_2/(1 + b_1^2 + b_2^2)$ is correct to $0(b_1 b_2)$; and in the case of (ii) of β_1 with $b_1(1 - b_2^2)/(1 + b_1^2 + b_2^2 + b_1^2 b_2^2)$ and β_2 with $b_2/(1 + b_1^2 + b_2^2 + b_1^2 b_2^2)$ is correct to $0(b_1^2 b_2)$.

```
•   X   •   X   •   X   •
X   •   X   •   X   •   X
•   X   •   X   •   X   •
X   •   X   •   X   •   X
```
Fig. 5

Coding methods   In the case of one-dimensional lattice models, we have noted that the two-sided nature of spatial models is not a serious problem. The absence of a simple *exact* method of converting to one-sided finite models in the case of two-dimensional models has been ingeniously by-passed by a method of coding introduced by Besag (1974a). Thus in the case of the scheme

$$E\{X_{rs} \mid \text{all other values}\} = \beta_1 (X_{r-1,s} + X_{r+1,s})$$
$$+ \beta_2 (X_{r,s-1} + X_{r,s+1})$$

we label the interior sites of the lattice alternately X and . , as shown in Fig. 5. This allows the values associated with the X sites (or alternatively the . sites) to be taken, conditional on the values at the . sites (or, in the alternative case, the X sites) as independent, and the likelihood function for these values is therefore simple to write down, and conditional regression analyses straightforward. With higher-order schemes involving longer-range dependences, the method is still applicable. Thus for second-order schemes a coding pattern as in Fig. 6 could be used, where the values at the X sites are considered conditional on the values at the . sites. This provides four sets of estimates by suitable choice of coding framework.

The efficiency of these coding methods is in general not known, but some information may be gained from the one-dimensional case. Here the corresponding coding pattern

```
•   X   •   X   •   X   •   X
•   •   •   •   •   •   •   •
•   X   •   X   •   X   •   X
•   •   •   •   •   •   •   •
```
Fig. 6

Table I

$\lvert \rho \rvert$	0.1	0.2	0.3	0.4	0.5	0.6	0.7	0.8	0.9
Efficiency { Single pattern	0.96	0.85	0.70	0.52	0.36	0.22	0.12	0.05	0.01
Average of both	0.99	0.96	0.91	0.84	0.75	0.64	0.51	0.36	0.19

for a first-order scheme was first in effect proposed by Ogawara (1951) and studied further by Hannan (1955). In that context the interest of the coding method was primarily to provide an exact, as distinct from an asymptotic, estimator and test; but from the present standpoint the interest is in the efficiency as well as the convenience of the method. Hannan gives two tables showing the efficiency of estimating the one regression or correlation coefficient ρ (i) by one coding pattern, (ii) by averaging the estimates from the two alternative patterns. His results are shown in Table I. It will be seen that the efficiency is high for low ρ, but not very great as ρ increases in the case of a single pattern, though considerably better for the average from both patterns than if one only is used.

Some recent results for the two-dimensional case by Besag and Moran (1975) are consistent with the expectation suggested by Table I. These, which are for the single pattern, might for reasonable comparability be plotted against the *multiple* correlation of a site value with its nearest neighbour values, and if then compared with the corresponding curve adapted from Table 1, show a similar behaviour, though with rather lower efficiency in the two-dimensional case.

2.1.1 Markov fields in continuous space

We may now return to the case of continuous space, at least for processes definable as limit processes first defined for lattice space and then allowing the grid dimensions to decrease (i.e. separable processes). If $X(\mathbf{r})$ is not to be related to *a priori* rectangular axes, it must not only be laterally symmetric but isotropic (in any case lateral symmetry can be converted to isotropy by choice of scale). It follows that such linear conditional nearest-neighbour models in continuous space must be normal, with spectral form (in two dimensions)

$$(\kappa^2 + \omega_x^2 + \omega_y^2)^{-1}, \tag{43}$$

in contrast to the 'autoregressive model'

$$\left(\frac{\partial^2}{\partial_x^2} + \frac{\partial^2}{\partial_y^2} - \kappa^2\right) X(\mathbf{r}) = Y(\mathbf{r}) \tag{44}$$

(where $Y(\mathbf{r})$ is an improper process approximated by a 'stationary' process over $\mathbf{r}$ with rapidly decreasing autocovariance), which has spectral density function proportional to

$$(\kappa^2 + \omega_x^2 + \omega_y^2)^{-2}. \tag{45}$$

The first form (1) may be derived from the linear spatial-temporal model

$$\left[\frac{\partial}{\partial t} + \kappa^2 - \left(\frac{\partial^2}{\partial_x^2} + \frac{\partial^2}{\partial_y^2}\right)\right] X(\mathbf{r}, t) = Y(\mathbf{r}, t) \tag{46}$$

as its stationary marginal spatial spectrum.

The identification of (43) and (45) as analogues of the results (25) and (7) respectively in §2.1 throws further light on their own classification as spatial models, following their original introduction by Whittle (1954; cf. also Heine, 1955; Whittle, 1962; Bartlett, 1966 §9.4; Moran, 1973).

2.2 Binary and other discrete variables

Finally, let us consider nearest-neighbour models for discrete variables, and in particular for the special but very important case of binary variables, taking the two values 0, 1, say (or alternatively, especially in the situation when each value is equally likely, the two values − 1 and 1, so that $E\{X\} = 0$).

The development and study of such models has a remarkable history which underlines the essential unity of statistical theory remarked on in the Preface. Perhaps I may trace some of my own steps in the pursuit of such models, in order to illustrate how the recognition of this unity in the present context has grown.

In 1967 (Bartlett, 1967a, Appendix II) I used a one-sided linear model to represent 'contagion' models on a lattice. This 'open-ended' process had the limitations already referred to when used as a model in what was strictly a symmetrical spatial context, but it nevertheless did represent a simple 'one-sided nearest-neighbour model' which enabled me to fit such models to non-random data (see also Part II). Their slight spatial asymmetry, however, led to the development of further one-sided approximations to the symmetrical model with spectrum (25), which could at least be realized by the spatial-temporal linear model (19) of §2.1.

At the same time I was interested to see if these models threw any light on the physical lattice problem familiar for many years as the Ising problem. Here it was customary to start with the classical simultaneous distribution for $X_{r,s}$ of the form

$$p(\mathbf{x}) = C \exp \left\{ - \alpha \sum_r x_{rs} - \gamma_1 \sum_{r,s} x_{rs} x_{r-1,s} \right.$$
$$\left. - \gamma_2 \sum_{r,s} x_{rs} x_{r,s-1} \right\} \tag{47}$$

determined by its required Gibbs form

$$p(\mathbf{x}) = C \exp \left(- H/\kappa T \right) \tag{48}$$

where H was the energy, in a physical context which is not relevant for our present purpose. I found, however, that no *linear* spatial-temporal nearest-neighbour binary model could give rise to the required distribution (47), and more awkward non-linear processes had to be introduced. In fact, we have seen from the work of J. E. Besag, noted in §2.1, that the linear spatial-temporal nearest-neighbour binary model is incompatible with *any* marginal spatial nearest-neighbour binary model.

2.2.1 The auto-logistic model

In particular, the spectrum (25) of §2.1 is not a possible one for such a binary model, and Besag (1972b) introduced the *auto-logistic* binary model to be compatible with the requirement (17) of §2.1. This model now becomes identical with (47) of §2.2, and is the *only possible binary model* with lateral symmetry. (Other workers (e.g. Spitzer, 1971; see Grimmett, 1972) had reached an equivalent conclusion.) This auto-logistic model, in terms of 0, 1 values, may be written

$$p(x \mid t,u,v,w) = \frac{\exp\{-x[\alpha + \gamma_1(t + u) + \gamma_2(v + w)]\}}{1 + \exp -\{\alpha + \gamma_1(t + u) + \gamma_2(v + w)\}}.$$

(49)

This identification of models leads to the unfortunate consequence that the correlational properties of the only possible binary model (49) with the required bilateral properties are most complicated. Even in the simplest case in equation (47) of §2.2 of $\alpha = 0$ (when $x = \pm 1$), $\gamma_1 = \gamma_2$, the correlations exhibit a singularity at a critical value of γ_1 (corresponding to a critical temperature in a physical context context), as was originally shown by Onsager (1944). Derivations of his solution for the nearest-neighbour correlation are still very involved; and formulae for longer-range correlations even more cumbersome. There are moreover further intriguing and relevant questions still not completely resolved, such as the extent to which the models for infinite lattices are uniquely defined (see, for example, Dobrushin, 1968 a,b,c; Hammersley, 1972).

However, in spite of some expedient by-passing of these correlational properties in fitting auto-logistic models to data (see Chapter 4), they are very important and further discussion of them is given below (cf. Bartlett, 1971b, 1972). The one-dimensional case is included, for this may be needed for some examples, and for non-zero α is by no means completely trivial. For

$$p(\mathbf{x}) = C \exp\{-\alpha\Sigma x_i - \gamma_1 \Sigma x_i u_i - \gamma_2 \Sigma x_i v_i\}$$

(50)

where for convenience the sites are labelled by i, and in the two-dimensional case (49) $x_i \equiv x_{r,s}$, $u_i \equiv x_{r-1,s}$, $v_i \equiv x_{r,s-1}$, (and in the one-dimensional case $x_i \equiv x_r$, $u_i \equiv x_{r-1}$, and v_i does not arise), consider

$$K(\theta, \phi_1, \phi_2) \equiv \log E \{\exp (\theta S_n + \phi_1 U_n + \phi_2 V_n)\} \quad (51)$$

where $S_n = \Sigma x_i$, $U_n = \Sigma x_i u_i$, $V_n = \Sigma x_i v_i$.

The suffix n denotes the number of sites, at present finite. The form of (50) ensures that

$$K(\theta, \phi_1, \phi_2) = \log C(\alpha, \gamma_1, \gamma_2)$$
$$- \log(\alpha - \theta, \gamma_1 - \phi_1, \gamma_2 - \phi_2) \quad (52)$$

whence $\quad K(\alpha, \gamma_1, \gamma_2) = \log C(\alpha, \gamma_1, \gamma_2) - \log C(0, 0, 0),$

$$(53)$$

$$E\{S_n\} = [\partial K(\theta, \phi_1, \phi_2)/\partial\theta]_{\theta,\phi_1,\phi_2=0}$$
$$= \partial \log C(\alpha, \gamma_1, \gamma_2)/\partial\alpha, \quad (54)$$

etc.

Note also that

$$\log C(\alpha, \gamma_1, \gamma_2) - \log C(0, 0, 0)$$
$$= - K_0(- \alpha, - \gamma_1, - \gamma_2), \quad (55)$$

where K_0 denotes K for $\alpha = \gamma_1 = \gamma_2 = 0$. Hence

$$m \equiv E\{S_n\} = \partial K_0(- \alpha, - \gamma_1, - \gamma_2)/\partial\alpha \quad (56)$$

etc., and may be alternatively derived from the properties of *purely random* lattice configurations. This is the method usually employed (although not always with an explanation of its logical basis). For example, in the one-dimensional problem with $x_i \equiv x_r = \pm 1$, $\gamma_1 \equiv \gamma$, K_0 may be derived by a standard matrix-powering technique (see Bartlett, 1966, §2.22). The relevant matrix is, if $\phi_1 \equiv \phi$,

$$M \stackrel{.}{=} \begin{pmatrix} \frac{1}{2}e^{\phi - \theta} & \frac{1}{2}e^{- \phi - \theta} \\ \frac{1}{2}e^{- \phi + \theta} & \frac{1}{2}e^{\phi + \theta} \end{pmatrix} \quad (57)$$

and M^n depends asymptotically for large n on the dominant eigenvalue

$$\lambda_1 = \tfrac{1}{2}e^\phi \{\cosh\theta + [\sinh^2\theta + e^{-4\phi}]^{\frac{1}{2}}\}, \tag{58}$$

giving

$$\log C(\alpha,\gamma) - \log C(0,0) \sim -n \log(\tfrac{1}{2}e^{-\gamma} \{\cosh\alpha + [\sinh^2\alpha + e^{4\gamma}]^{\frac{1}{2}}\} \tag{59}$$

whence

$$m/n = \frac{\partial \log C(\alpha,\gamma)}{\partial\alpha} = \frac{-\sinh\alpha}{[\sinh^2\alpha + e^{4\gamma}]^{\frac{1}{2}}}$$
$$(=0 \text{ when } \alpha = 0), \tag{60}$$

$$m_2/n = \frac{\partial \log C(\alpha,\gamma)}{\partial\gamma}$$
$$= 1 - \frac{2e^{4\gamma}}{[\sinh^2\alpha + e^{4\gamma}]^{\frac{1}{2}} \{\cosh\alpha + [\sinh^2\alpha + e^{4\gamma}]^{\frac{1}{2}}\}}. \tag{61}$$

The nearest-neighbour *correlation* (in the usual statistical sense) is

$$r_1 = \frac{m_2/n - (m/n)^2}{1 - (m/n)^2} = \frac{\cosh\alpha - [\sinh^2\alpha + e^{4\gamma}]^{\frac{1}{2}}}{\cosh\alpha + [\sinh^2\alpha + e^{4\gamma}]^{\frac{1}{2}}}. \tag{62}$$

reducing to $-\tanh\gamma$ when $\alpha = 0$.

The derivation of r_1 even for $\alpha \neq 0$ by the above method is rather deceptively simple, as the method, although extended by Onsager in the two-dimensional case (for $\alpha = 0$), then involves much more complicated matrices. Even in the one-dimensional case notice that only the first-order correlation r_1 has been derived.

For an alternative combinatorial approach, suggested by B. L. van der Waerden and developed further by M. Kac and J. C. Ward (see Newell, G. F. and Montroll, E. W., 1953), consider the expression (in the two-dimensional case)

$$p(\mathbf{x}) = C\Pi_i \exp(-\alpha x_i) \cdot \exp(-\gamma_1 x_i u_i) \cdot \exp(-\gamma_2 x_i v_i). \tag{63}$$

Now

$$e^{-\alpha x_i} = \cosh(\alpha x_i) - \sinh(\alpha x_i) = \cosh \alpha \, (1 - x_i t_1),$$

where $t_1 = \tanh \alpha$, and

$$e^{-\gamma_1 x_i y_i} = \cosh \gamma_1 \, (1 - x_i u_i \Gamma_1), \text{ etc.}$$

where $\Gamma_1 = \tanh \gamma_1$. Hence

$$p(\mathbf{x}) = C \cosh^n \alpha \, \cosh^n \gamma_1 \, \cosh^n \gamma_2 \Pi_i (1 - x_i t_1)$$
$$\times \ (1 - x_i u_i \Gamma_1)(1 - x_i v_i \Gamma_2) \qquad (64)$$

Since some factors in x_i cancel out (because $x_i^2 = 1$) when summed over all x_i, this gives

$$1 = C \cosh^n \alpha \, \cosh^n \gamma_1 \, \cosh^n \gamma_2 \, S_0,$$

where S_0 is the constant term in the expansion of the product on the right. For example, in the one-dimensional case with $\alpha = 0$, this gives

$$1 = C \cosh^n \gamma \times \text{constant term in } \Pi_r (1 - x_r x_{r-1} \Gamma) \quad (65)$$

which leads obviously (for circular end-conditions) to

$$1 = C \cosh^n \gamma [1 + \tanh^n (-\gamma)],$$

or

$$C = 1/[\cosh^n \gamma + \sinh^n (-\gamma)] \sim \cosh^{-n} \gamma, \qquad (66)$$

whence

$$m_2/n = \rho_1 = -\tanh \gamma = -\Gamma,$$

as already deduced. Notice also that by multiplying each side of (64) above by $x_r x_{r-1}$ before summing, we can obtain $E\{x_r x_{r-1}\}$, and this gives ($\alpha = 0$)

$$\rho_1 = C \cosh^n \gamma \tanh(-\gamma),$$

whence

$$\rho_1 = \tanh(-\gamma), \text{ as before.}$$

In the two-dimensional case the method again becomes very involved, but it was shown by Kac and Ward that for

$\alpha = 0$ and an $n = N \times M$ rectangular lattice (M even)

$$C \sim \cosh^n \gamma_1 \cosh^n \gamma_2 \prod_{r=1}^{N} \prod_{s=1}^{\frac{1}{2}M} \left\{ (1 + x^2)(1 + y^2) - 2y \right.$$

$$\left. \times (1 - x^2) \cos \frac{2\pi r}{n} - 2x(1 - y^2) \cos \frac{2\pi s}{n} \right\} \qquad (67)$$

where $x = -\tanh \gamma_1$, $y = \tanh \gamma_2$. As $N, M \to \infty$, this leads to Onsager's result

$$[\log C]/n \to \log 2 + \frac{1}{2\pi^2} \int_0^\pi \int_0^\pi \log[\cosh 2\gamma_1 \cosh 2\gamma_2$$

$$+ (\sinh 2\gamma_1)(\cos \omega_1) + \sinh(2\gamma_2)(\cos \omega_2)] \, d\omega_1 \, d\omega_2 \qquad (68)$$

Differentiation of (68) with respect to γ_1 or γ_2 leads to formulae for the nearest-neighbour product-moments ρ_{10}, ρ_{01}. In the completely symmetric case with $\gamma_1 = \gamma_2 = \gamma$, the formula for $\rho_{10} = \rho_{01}$ may be put in the form (Bartlett, 1972)

$$\rho_{10} = -\frac{(\epsilon + \sqrt{(1 - \epsilon^2)}\beta_{10}(\xi)}{1 - 2\xi\overset{\circ}{\rho}_{10}(\xi)} \qquad (69)$$

where $\xi = -\epsilon\sqrt{(1 - \epsilon^2)}$, $\epsilon = \tanh 2\gamma$, and $\overset{\circ}{\rho}_{10}(2\beta_1)$ is the nearest-neighbour correlation for the symmetric $(\beta_1 = \beta_2)$ 'linear spectrum' (25) of §2.1. A numerical table of ρ_{10} in (69) is given in Table AII. Table AI gives values of $\overset{\circ}{\rho}_{rs}(2\beta_1)$.

Yet another possible approach (Bartlett, 1971b, 1972) is an extension of the spatial-temporal models of §2.1 to non-linear processes leading to the equilibrium distribution (50). The Markov process in time, defined by

$$P\{x_{i, t+dt} = x_{i, t} \mid \mathbf{x}_t\} = 1 - \lambda dt G_i(\mathbf{x}_t), \qquad (70)$$

is considered, where $\mathbf{x}_t$ denotes the n values of x_i simultaneously at time t, and $G_i(\mathbf{x}_t)$ is an arbitrary function in all the $x_{j, t}$ affecting $x_{i, t+dt}$. When the possible values of x_i are restricted to ± 1, equation (70) implies that any equilibrium distribution must satisfy

$$\sum_i G_i(\mathbf{x}) = \sum_i G_i(\mathbf{y}_i)p(\mathbf{y}_i)/p(\mathbf{x}), \tag{71}$$

where $\mathbf{y}_i \equiv \mathbf{x}$ except that $-x_i$ replaces x_i. For the distribution (50) to be possible, we find

$$\sum_i G_i(\mathbf{x}) = \sum_i G_i(\mathbf{y}_i) \exp\{2\alpha x_i + 2\gamma_1 x_i u_i' + 2\gamma_2 x_i v_i'\}, \tag{72}$$

where u_i', v_i' are the symmetrized forms of u_i, v_i (e.g. in one dimension, where $u_i \equiv x_{r-1}$, $u_i' \equiv \frac{1}{2}(x_{r-1} + x_{r+1})$). Equation (72) may most obviously be satisfied by the *local* (symmetric) solutions

$$G_i(\mathbf{x}) = \exp\{\alpha x_i + \gamma_1 x_i u_i' + \gamma_2 x_i v_i'\} f(\alpha x_i, \gamma_1 x_i u_i', \gamma_2 x_i v_i') \tag{73}$$

where f is some appropriate function, *even* in all its variables. It may (see, however, Bartlett, *loc. cit*; Besag, 1972) sometimes also be satisfied by *global* (one-sided) solutions

$$G_i(\mathbf{x}) = \exp\{\alpha x_i + \gamma_1 x_i u_i + \gamma_2 x_i v_i\} g(\alpha x_i), \tag{74}$$

but we consider here only the type (73). Possible choices of this type are, for $\alpha = 0$,

One dimension:

$$G_i(\mathbf{x}) = 1 + \epsilon u_r', \tag{75}$$

where $\epsilon \equiv \tanh(2\gamma)$

Two dimensions $(\gamma_1 = \gamma_2 = \gamma)$:

$$G_i(\mathbf{x}) = 1 + \frac{\epsilon}{1 + \epsilon^2} x_{rs} z_{rs}' - \frac{\epsilon^3}{z(1 + \epsilon^2)} x_{rs} z_{rs}'[(z_{rs}')^2 - 4], \tag{76}$$

where $z_{rs}' = u_{rs}' + v_{rs}'$. When $\alpha \neq 0$, we have

$$G_i(\mathbf{x}) = (1 + x_i \tanh \alpha) G_i(\mathbf{x}; \alpha = 0). \tag{77}$$

If we expand $p(\mathbf{x})$ by the unique orthogonal series

$$2^n p(\mathbf{x}) = 1 + \sum_i x_i \rho_i + \sum_{i,j} x_i x_j \rho_{ij} + \dots \tag{78}$$

where $\rho_i = E\{x_i\}$, $\rho_{ij} = E\{x_i x_j\}$, etc., we find that
term-by-term solution of (71) leads to the equation

$$x_i\{A_i(\mathbf{x})b(\mathbf{x}) + B_i(\mathbf{x})a(\mathbf{x})\} = 0, \tag{79}$$

where $G_i(\mathbf{x}) = A_i(\mathbf{x}) + x_i B_i(\mathbf{x})$, $p(\mathbf{x}) = a(\mathbf{x}) + x_i b(\mathbf{x})$, the
functions A, B, a, b not containing x_i explicitly. Thus for
$G_i(\mathbf{x})$ in (75), when $p(\mathbf{x})$ is symmetric so that $\rho_i = \rho_{ijk} = \ldots$
$= 0$, the coefficient of $x_i x_j$ leads to the equation

$$\rho_s + \tfrac{1}{2}\epsilon(\rho_{s-1} + \rho_{s+1}) = 0, \tag{80}$$

equivalent to the previous solution $\rho_1 = -\tanh\gamma$, and also to
$\rho_s = \rho_1^s$. For the one-dimensional case and $\alpha \neq 0$, the reader
may like to investigate the analogous equations and try to
derive results (60) and (62) from them. (Although the
equations now involve higher moments like ρ_{ijk}, these may
be reduced in the one-dimensional case, by using Markov
chain relations available in that case, to recurrence equations
for ρ_s of the type (80), thus providing solutions for ρ_s for all s.

In the two-dimensional case, even for $\alpha = 0$ when the exact
solution (69) for ρ_1 is known, this last approach has not so
far yielded exact solutions, but an approximate solution for
ρ_1 based on a quasilinearization of the equations for ρ_s has
been investigated (for further details, see Bartlett, 1971b, 1972,
1974). It appears quite accurate *below* the critical value of γ
when compared with the exact solution, and has also been
evaluated in the three-dimensional case ($\alpha = 0$, $\gamma_1 = \gamma_2 = \gamma_3 = \gamma$). It could if required be extended in the two- (and
three-) dimensional cases for $\alpha \neq 0$.

2.2.2 Sub-critical 'temperatures'

The discussion of the auto-logistic model in §2.2.1 does not
treat in any detail the situation for high γ(and $\alpha = 0$),
corresponding to a low temperature T in a physical context,
below the critical value T_c. Conditions in this range of γ are
perhaps less likely to be of interest in biological applications,
but in view of their relevance in physical applications some
further remarks when γ is in this range are made below. First

it should be noted that the distribution (50) and the expansion (78) of §2.2.1 are defined for finite n, and critical behaviour only strictly arises in the limit as $n \to \infty$ (cf. (67) and (68)). This means that the interpretation of product moments arising from (79) as spatially analogous to terms in an *infinite* time series must be made with caution, as *non-ergodic* limiting components could complicate the passage to the limit.

Suppose, for example, we consider the one-dimensional case with non-zero mean m ($\alpha \neq 0$). The equation for m leads to only one solution, which vanishes as $\alpha \to 0$. However, in the higher-dimensional cases this is no longer true, so that under some conditions the equations for m may provide a solution $m(\alpha)$ such that $\lim_{\alpha \to 0} m(\alpha) \neq 0$. In a physical context this is proportional to the *spontaneous magnetization*, but is clearly a property of the auto-logistic model as such. It indicates that we may put, when $n = \infty$,

$$x_i = y_i + m, \tag{81}$$

even in the $\alpha = 0$ case, where m is a non-ergodic or long-range component, such that

$$E\{x_i x_j\} = E\{y_i y_j\} + m^2, \tag{82}$$

where $E\{y_i\} = 0$. $E\{x_i\}$ is also zero, as for the symmetric situation $\alpha = 0$ we have strictly $E\{x_i\} = \frac{1}{2}m + \frac{1}{2}(-m) = 0$, only $|m|$ or m^2 being of relevance in (82); however, the situation is exactly *in any single realization* like (81), and the *full* expansion (78) should be retained even for $\alpha = 0$.

This model for infinite n may be checked from the situation for high γ by considering the equations near $\rho_{10} = 1$. Thus the complete equations may be written (for $t_1 = 0$)

$$x_i \{ (1 + \epsilon^2 s_i)(m + {\sum_j}' x_j \rho_{ij} + {\sum_{jk}}' x_j x_k \rho_{ijk} + \dots)$$

$$+ \epsilon z_i'(1 + m {\sum}' x_j + {\sum}' x_j x_k \rho_{jk} + \dots) \} = 0 \tag{83}$$

where $z_i' = \frac{1}{2}(x_i' + x_i'' + x_i''' + x_i'''')$, $x_i', x_i'', x_i''', x_i''''$ being the

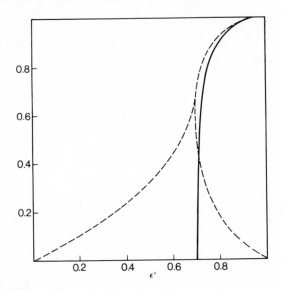

Fig. 7 *Structure of nearest-neighbour correlation for the two-dimensional Ising model. Upper dotted curve:* ρ_{10}. *Lower dotted curve:* r_{10} *(identical with* ρ_{10} *below* $\epsilon' = 1/\sqrt{2}$*). Solid curve:* m^2 *(From J. Appl. Prob., 1974b)*

four nearest-neighbours of x_i, and s_i the mean product of pairs $x_i'x_i''$, etc. The summation $\sum'_j$, excludes i, and $\sum'_{jk}$, etc. similar summation over $j < k < \ldots$. The first equation (coefficient of x_i) is

$$(1 + 2c)m + \epsilon^2 P_{10}(2) = 0 \tag{84}$$

where $P_{10}(2) = E\{x_i s_i\}$. As a *first* approximation, we may write $P_{10}(2) \sim m^3$, whence a possible solution near $m^2 = 1$ is

$$m^2 \sim (1 + 2\epsilon)/\epsilon^2 = 1 - y^2 - 2y^3 + 0(y^4) \tag{85}$$

where $\epsilon' = -\epsilon = 1 - y$. This may be compared with the exact formula for m^2 below T_c due to Yang (1952)

$$m^2 = [1 - \sinh^{-4} 2\beta]^{\frac{1}{4}} = 1 - y^2 - 3y^3 + 0(y^4),$$
$$(\beta = -\gamma). \tag{86}$$

Agreement to the next order may be obtained by a second approximation in which we write in (84) and further relations such approximations as

$$E\{x_i x_i' x_i''\} \sim m^3 + m(2\rho_{10} + \rho_{11}) \tag{87}$$

(see Bartlett, 1974b).

It is of interest to compare the behaviour of the correlation

$$r_{10} = (\rho_{10} - m^2)/(1 - m^2) \tag{88}$$

with the one-dimensional correlation r_1 in (62). The latter is easily shown to decrease with increasing m, and similarly r_{10} *decreases* below T_c as m^2 increases (see Fig. 7).

2.3 General specification of conditional lattice systems

We come now to the more general formulation (Besag, 1974a) mentioned earlier, defining n sites i and neighbours affecting x_i, in a more arbitrary manner. It is supposed that the conditional probability distribution. $P\{x_i \mid$ all other site values$\}$, is specified and positive for all i, and that for the possible (finite set of) values of x_i no restrictions hold for the simultaneous realizations. Then for two given realizations x and y

$$p(\mathbf{x}) = p(x_n \mid x_1, \ldots x_{n-1}) p(x_1, \ldots x_{n-1})$$

$$= \frac{p(x_n \mid x_1, \ldots x_{n-1}) p(x_1, \ldots x_{n-1}, y_n)}{p(y_n \mid x_1, \ldots x_{n-1})}$$

$$= \frac{p(x_n \mid x_1, \ldots x_{n-1})}{p(y_n \mid x_1, \ldots x_{n-1})} \frac{p(x_{n-1} \mid x_1, \ldots x_{n-2}, y_n)}{p(y_{n-1} \mid x_1, \ldots x_{n-2}, y_n)}$$

$$\times p(x_1, \ldots x_{n-2}, y_{n-1}, y_n)$$

$$= \ldots = p(\mathbf{y}) \prod_{i=1}^{n} \frac{p(x_i \mid x_1, \ldots x_{i-1}, \ldots y_n)}{p(y_i \mid x_1, \ldots x_{i-1}, y_{i+1}, \ldots y_n)}. \tag{89}$$

Next we define the set of neighbours of site i by the condition $p(x_i \mid$ all other values) is dependent on x_j if and only if site j ($\neq i$) is a 'neighbour' of site i. Any set of sites consisting of a single site i, or in which every site is a neighbour of every other site in the set is called a 'clique'. For convenience and without loss of generality, we still suppose that 0 is one possible value of x, and define

$$L(\mathbf{x}) = \log p(\mathbf{x}) - \log p(\mathbf{0}). \tag{90}$$

Expand $L(\mathbf{x})$ in the unique expansion below,

$$L(\mathbf{x}) = \Sigma_i x_i G_i(x_i) + \Sigma_{i,j} x_i x_j G_{i,j}(x_i, x_j) + \ldots +$$
$$x_1 x_2 \ldots x_n G_{1,2,\ldots n}(x_1, x_2, \ldots x_n) \tag{91}$$

where in the summations no $i, j, \ldots$ are repeated. (For example,

$$x_i x_j G_{i,j}(x_i, x_j) = \log p(0, \ldots 0, x_i, 0, \ldots 0, x_j, 0 \ldots 0)$$
$$-\log p(0, \ldots 0, x_i, 0, \ldots 0) - \log p(0, \ldots 0, x_j, 0, \ldots 0)$$
$$+\log p(\mathbf{0}).)$$

Then it will be shown (the Hammersley-Clifford theorem) that the functions $G_{i,j,\ldots s}$ in (91) are arbitrary, except that they may be non-null if and only if the sites $i, j, \ldots s$ form a clique.

Note first that from (89), if $\mathbf{x}_i$ denotes $(x_1, \ldots x_{i-1}, 0, x_{i+1} \ldots x_n)$ then

$$\exp\{L(\mathbf{x}) - L(\mathbf{x}_i)\} = p(x_i \mid x_1, \ldots x_{i-1}, x_{i+1}, \ldots x_n)/$$
$$p(0 \mid x_1, \ldots x_{i-1}, x_{i+1}, \ldots x_n)$$

and can only depend on x_i and values at 'neighbouring' sites. Consider in particular site 1. We have

$$L(\mathbf{x}) - L(\mathbf{x}_1) = x_1 \{G_1(x_1) + \Sigma'_j x_j G_{1,j}(x_1, x_j) + \ldots\}$$

where the dashed summations do not involve site 1. Suppose site k ($\neq 1$) is not a neighbour of site 1. The $L(\mathbf{x}) - L(\mathbf{x}_1)$ must be independent of x_k for all $\mathbf{x}$. Putting $x_i = 0$ for $i \neq 0$ or k, we see that $G_{1,k}(x_1, x_k) = 0$. Similarly the higher-order G functions involving x_1 and x_k must be null. Similar results also follow for any pair of sites that are not neighbours, and the theorem follows.

The theorem may be extended to more general x variates provided of course that the sum or integral of exp $L(\mathbf{x})$ remains finite.

In the cases where $L(\mathbf{x})$ depends only on contributions from cliques containing no more than two sites, the expansion for $L(\mathbf{x})$ must be of the form

$$L(\mathbf{x}) = \Sigma_i x_i G_i(x_i) + \Sigma_{ij} x_i x_j G_{i,j}(x_i, x_j) \ . \tag{92}$$

In the further special case where x is binary $(0,1)$, $G_i(x_i)$ and $G_{i,j}(x_i, x_j)$ must necessarily be constants for each i, j. If the system is also 'stationary', G_i is an absolute constant, and $G_{i,j}$ depends only on the (signless) vector displacement of i and j. The simplest form of the auto-logistic model so defined is the one previously discussed, where only the *nearest*-neighbours on a rectangular lattice contribute. The auto-logistic model for a regular triangular lattice, where each site has *six* nearest neighbours in a plane, is similarly readily obtained.

It is recalled that the number n of sites in the above discussion is finite. This is strictly consistent with practical situations, but the latter must then be considered for terminal boundary conditions (i.e. we cannot have strict 'stationarity'); alternatively, if n increases indefinitely the further problems of ergodicity previously mentioned will arise.

2.3.1 Spatial-temporal processes for other auto-schemes

Other auto-models of the conditional nearest-neighbour or Markov Field type, while necessarily conforming to the general form (92) of §2.3, may be defined even on the rectangular lattice if the basic distribution is not restricted to the binary or normal. They may (cf. Besag, 1974b) be conveniently generated by relaxing the restriction in our spatial-temporal process for the auto-logistic model from binary $(0,1)$ variables to positive integer variables. Thus in place of equation (70) of §2.2.1 we may write

$$P\{x_{i,t+dt} = y \mid \mathbf{x}_t\} = \lambda G_i(y, \mathbf{x}_t)dt, \ (y \neq x_{i,t}). \tag{93}$$

The limiting equilibrium distribution $P(\mathbf{x})$ must satisfy (in place of equation (71) of §2.2.1)

$$\Sigma_i S_y p(\mathbf{y}_i)G_i(x, \mathbf{y}_i) = p(\mathbf{x})\Sigma_i S_y G_i(y, \mathbf{x}), \tag{94}$$

where S_y denotes summation over values of $y \neq x$ at site i, and $\mathbf{y}_i = \mathbf{x}$ except that the value at site i is not x but y. We choose the convenient term-by-term or 'local' solution of (94) viz

$$p(\mathbf{y}_i)G_i(x, \mathbf{y}_i) = p(\mathbf{x})G_i(y, \mathbf{x}). \tag{95}$$

Auto-binomial model   Suppose in particular that x_i only changes at most by one, and in particular

$$G_i(y, \mathbf{x}) = \begin{cases} (n - x_i)\gamma(\mathbf{x}) & \text{for} \quad y = x_i + 1 \\ x_i \delta_i(\mathbf{x}) & \text{for} \quad y = x_i - 1 \end{cases}$$

where

$$\begin{cases} \gamma(\mathbf{x}) = \exp[\alpha + \Sigma_j \beta_j x_j] \\ \delta(\mathbf{x}) = \exp[\alpha' + \Sigma_j \beta'_j x_j], \end{cases}$$

x_j being the nearest neighbours of x_i. If we write

$$\beta(\mathbf{x}) = \exp[\alpha - \alpha' + \Sigma_j(\beta_j - \beta'_j)x_j]$$

where $\beta_j - \beta'_j$ is assumed to have at least lateral symmetry in relation to x_i and x_j, then the limiting conditional distribution at site i is binomial with parameters n and $\theta = \beta(\mathbf{x})/[1 + \beta(\mathbf{x})]$. When $n = 1$, this model reduces to the auto-logistic $(0, 1)$ model.

Auto-Poisson model   If alternatively

$$G_i(y, \mathbf{x}) = \begin{cases} \gamma(\mathbf{x}) & \text{for} \quad y = x_i + 1 \ (\text{except if } x_i = 0) \\ x_i \delta(\mathbf{x}) & \text{for} \quad y = x_i - 1 \end{cases}$$

then the limiting conditional distribution at site i is Poisson with mean $\beta(x)$. At first sight this appears a potentially useful model, but the requirement that $p(x)$ is a valid distribution imposes the rather severe restriction that $\beta_j - \beta_j' \leq 0$ (Besag, 1974a), implying competition rather than positive contagion between neighbouring sites.

PART II

Examples of Statistical Analyses

CHAPTER THREE

Analyses of Continuous and Point Processes

3.1 Continuous processes $X(\mathbf{r})$. Pielou's example

Examples of processes as defined in §1.1 are not difficult to imagine, but not strictly very numerous, as it is more customary to consider either extracted lattice processes X_{rs} (as in one dimension), or derived lattice processes Y_{rs}, where $Y(\mathbf{r})$ was defined in equation (1) of §1.1.

Such examples will be deferred until §4.1, and this section will deal with the special problem mentioned in the Preface, of a mosaic pattern where $X(\mathbf{r})$ is either 1 or 0 (presence or absence). Such an example was cited by Pielou (1964), and analysed on the basis of a Markov chain transect model, that is, it was assumed that on any straight line intersecting the area, the probability of switching to vegetation if there were none at the moment was, say, λ_1 ds in a distance ds, and λ_2 ds of a switch to no vegetation if there were vegetation present. This model implies exponential intervals of vegetation of mean length $1/\lambda_2$ and exponential intervals of no vegetation of mean length $1/\lambda_1$. The stochastic equation for the 0,1 variable X_s is, moreover,

$$dX_s = -\lambda_2 X_s dt + \lambda_1 (1 - X_s)dt + dZ_s, \tag{1}$$

implying an exponential correlation function of $\lambda_1 + \lambda_2$, viz

$$\rho_s = e^{-(\lambda_1 + \lambda_2)s}, \quad (s > 0). \tag{2}$$

It was not immediately clear that such a model was appropriate or even admissible in such a two-dimensional

Table II Observations on a population of *Antennaria umbrinella* (from Pielou, 1964, Table 1)

Event	$s = 1$	2	3	4	5	6	Total
00	135	133	124	114	129	119	$[0 = 00 + \frac{1}{2}(01)]$
01 (10)	20	28	43	59	44	54	878
11	45	39	33	27	27	27	$[1 = 11 + \frac{1}{2}(01)]$ 322
Total	200	200	200	200	200	200	1200

context (it may, however, be shown that it is consistent with vegetation areas bounded by random straight lines; see Switzer, 1965) but the correlation function (2) does in fact fit the observed correlations reasonably well (Bartlett, 1964a). More natural correlation functions in this two-dimensional context (though not perhaps for a 0,1 variable) would correspond to the spectral functions (43) or (45) of §2.1.1.

The sampling data, which are given in full in Table II, were obtained from quadrat-pairs, each quadrat being a circle of 2 cm diameter, and the distance between the pair being in units of 2 cm. Of the two degrees of freedom for each quadrat-pair type, one merely checks the randomness of the sampling, and the other corresponds to the correlation-coefficient value given in Table III.

Table III *Pielou's transect example (Antennaria umbrinella)*

Interval	Observed correlation	Exponential
2 cm	0.75	0.78
4 cm	0.64	0.61
6 cm	0.46	0.48
8 cm	0.27	0.37
10 cm	0.41	0.29
12 cm	0.32 (±0.07)	0.23

3.2 Point processes. A simulated clustering model.
χ^2 analyses. Mead's randomization test

Examples of two-dimensional point-processes $N(\mathbf{r})$, with
spectral and other analyses, have been given in Bartlett
(1964b). In that paper I noted the value of a simple but
powerful analysis of variance of the counts in a suitable grid,
rather on the lines advocated by Greig-Smith (1957), though
not necessarily so systematic. If a preliminary check of non-
randomness is required, a chi-square analysis of the cell
counts, possibly after extracting rows and columns, may
prove informative.

The examples given, which were not ideal, included the
analysis of data from Numata (1961) on 'stands' of Japanese
black pine saplings, and also one of hypothetical data
conforming to a clustering model (offspring only) with
Poisson size families and isotropic Gaussian dispersal. For the
simulated data shown in Fig. 1 (§1.2.1) of 100 points with
Poisson family size 2, and dispersal standard deviation 1 (in
each dimension), the clustering in this hypothetical example

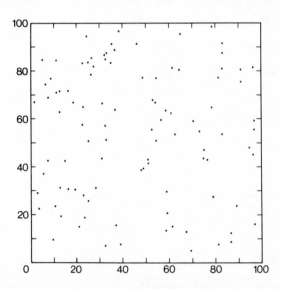

Fig. 8 *Clustering model (cf. Fig. 1, which is same realization with
one-fifth the dispersion). (From Adv. Appl. Prob, 1974a)*

Table IVa (*based on original simulation*)

Counts					Chi-square analysis		
9	1	4	0	4	Rows	8.1	(4)
3	1	4	2	5	Columns	4.7	(4)
1	6	3	4	0	Remainder	34.2**	(16)
12	5	5	4	3			
3	8	2	6	5	Total	47.0**	(24)

is very marked, more so than appears in a 5 × 5 cell count, although the heterogeneity of these may be demonstrated (Table IV). The spectral function for this is given by

$$g(\omega) = \lambda(1 + 2e^{-(\omega_1^2 + \omega_2^2)}). \tag{3}$$

In later discussion of this example, I have pointed out the dangers of the rather arbitrary scale in an analysis like Table IV, as distinct from Greig-Smith's more systematic counting or a full spectral analysis. As an illustration I took the same hypothetical example, but blew up the dispersal scale five-fold, so that the spectral function was now

$$g(\omega) = \lambda(1 + 2e^{-25(\omega_1^2 + \omega_2^2)}); \tag{4}$$

the *same* simulated points were used otherwise, apart from any necessary adjustments round the margins. Rather

Table IVb (*based on modified simulation*)

Counts					Chi-square analysis		
5	2	5	1	4	Rows	6.3	(4)
4	2	6	1	3	Columns	3.5	(4)
1	5	5	4	1	Remainder	31.7*	(16)
12	4	5	3	5			
2	9	2	7	2	Total	41.5*	(24)

* Denotes $P = 0.05$ significance.
** 0.01 significance.
*** 0.001 significance.

remarkably, although the simulated data shows much less visual evidence of clustering (Fig. 8; the cases of very near nearest-neighbours are fortuitous and do not even come from the same families) the cell counts and corresponding chi-square analysis tell very much the same story as before (Table IVb).

This similarity of the two analyses is, however, rather dependent on a particular choice of grid, as may be illustrated by carrying through a fuller χ^2 analysis of the Greig-Smith nested type. It is convenient for this analysis to choose a grid of the $2^k \times 2^k$ type, and the total square for each simulation was divided into 16×16 units, the counts for these being shown in Tables Va and Vb, which also show the corresponding χ^2 analyses. The second simulation does not now succeed in demonstrating significance at all; and from the theoretical mean squares shown, which are available as we know the underlying model (see below), it is clear that the clustering is too diffuse to show up in this analysis. This is somewhat unlucky; it will be seen that the actual mean squares are well below expection for the larger squares. It may be checked that this wide variation from theory is not abnormal, for the lower 0.05 limits on the true expectations are even further below the observed mean squares (even on the basis of standard significance levels, which would tend to underestimate the variability to be expected from a heterogeneous or clustering pattern).

The theoretical mean squares are computed as follows. For a stationary point process $dN(\mathbf{r})$ (cf. § § 1.2, 1.21) we have

$$E\{N^2\} = \int E\{dN(\mathbf{r})\} + \int\int E\{dN(\mathbf{r})dN(\mathbf{s})\}, \qquad (5)$$

or

$$\sigma^2(N) = \lambda A + \int\int w(\mathbf{r} - \mathbf{s})d\mathbf{r}d\mathbf{s}, \qquad (6)$$

where N is the total count over a square of area A, and the integration of both $\mathbf{r}$ and $\mathbf{s}$ in the last integral is over this region. If $w(\mathbf{r} - \mathbf{s})$ is of the form $Cw_1(x_1 - x_2)w_1(y_1 - y_2)$, where $\mathbf{r} = (x_1, y_1)$, $\mathbf{s} = (x_2, y_2)$, then this integral may be

Table Va *Simulation 1*

	(1)	(2)	(3)	(4)	(5)	(6)	(7)	(8)	(9)	(10)	(11)	(12)	(13)	(14)	(15)	(16)	Total
(1)	1	.	1	.	.	.	.	.	1	1	.	.	.	.	.	.	4
(2)	.	.	2	1	.	.	.	.	.	.	.	.	.	.	3	.	6
(3)	.	3	1	.	.	.	3	.	1	.	.	.	.	.	.	.	8
(4)	.	.	.	.	.	.	1	.	.	1	.	1	.	.	.	.	3
(5)	.	.	.	.	.	.	.	.	.	.	1	.	.	.	.	1	2
(6)	.	1	2	.	.	.	.	.	.	.	.	.	.	3	.	.	6
(7)	.	.	.	.	.	2	4	.	.	.	.	.	.	2	.	.	8
(8)	.	.	.	2	.	.	.	.	3	4	.	.	.	.	.	.	9
(9)	.	.	.	.	.	.	.	.	.	.	.	.	.	.	.	.	0
(10)	.	2	.	.	.	.	.	.	.	.	.	.	.	.	.	.	2
(11)	.	2	3	.	.	.	3	.	.	.	.	.	.	.	2	.	10
(12)	.	.	6	.	1	.	3	.	.	.	.	.	.	.	.	.	10
(13)	.	.	.	.	.	.	3	.	.	.	3	2	1	.	.	.	9
(14)	1	.	.	1	4	.	.	.	1	1	1	.	.	.	.	1	10
(15)	.	.	2	.	.	.	.	.	.	.	1	.	.	2	1	1	7
(16)	.	.	.	.	3	.	.	.	.	.	3	.	.	.	.	.	6
Total	2	8	17	4	8	2	17	0	6	7	9	3	1	7	6	3	100

χ^2 *analysis*

Item	D.F.	S.S.	M.S.	Theory
Within 4's	192	419.84***	2.19	2.12
Between 4's within 16's	48	115.84***	2.41	3.00
Between 16's within 64's	12	25.28*	2.11	3.00
Between 64's	3	4.64	1.55	3.00
Total	225	565.60***	2.22	2.34

*Denotes $P = 0.05$ significance.   **0.01 significance.   ***0.001 significance.

Table Vb *Simulation 2*

	(1)	(2)	(3)	(4)	(5)	(6)	(7)	(8)	(9)	(10)	(11)	(12)	(13)	(14)	(15)	(16)	Total
(1)			1				1	1	2					1			6
(2)			1	1									1				3
(3)		2	1					1							2		6
(4)	1			1		1				1		1			1		6
(5)							1	1	2								4
(6)			1					1						1		1	4
(7)			1			2		1				1	1	2			8
(8)	1			1	1	1	2			2							8
(9)		1		1				1		1							4
(10)						1									1		2
(11)	1	3	2			1	1							1	1		10
(12)			3	1					1	1	1						7
(13)	1		2			1	1	1				2	1	1			10
(14)					1	1				1	1	1					5
(15)			1	1	3	1				1	1		2		1		11
(16)			1	1		1				1		1	1				6
Total	4	6	14	7	5	10	6	7	5	8	3	6	6	6	6	1	100

χ^2 *analysis*

Item	D.F.	S.S.	M.S.	Theory
Within 4's	192	193.28	1.01	1.08
Between 4's within 16's	48	57.28	1.19	1.45
Between 16's within 64's	12	13.68	1.14	1.93
Between 64's	3	4.40	1.47	3.00
Total	225	268.64	1.05	1.21

written

$$C\left[\int_{-h}^{h}(h-|u|)w_1(u)du\right]^2 = CI^2,$$ (7)

say, where $A = h^2$. In particular, for the clustering model used for these simulations, for which

$$w(\mathbf{r}) = \lambda m_0 \exp\{-(x_1^2 + y_1^2)/4\sigma^2\}/(4\pi\sigma^2),$$ (8)

we have $C = \lambda m_0$ and

$$w_1(u) = (4\pi\sigma^2)^{-\frac{1}{2}}\exp(-\frac{1}{2}u^2/\sigma^2).$$ (9)

The variance-mean ratio becomes $1 + m_0 I^2/A = 1 + m_0 J^2$, where

$$J = 2\int_0^h (1-u/h)(4\pi\sigma^2)^{-\frac{1}{2}}\exp\{-u^2/(4\sigma^2)\}du.$$ (10)

Notice that for σ small compared with h, this ratio tends to $1 + m_0$, or 3 in the simulations, for which $m_0 = 2$. For σ large, $J \to \frac{1}{2}h\sqrt{/(\pi\sigma^2)}$, and the ratio tends to $1 + h^2(2\pi\sigma^2)$. Even for Simulation 2, the smallest value of h is $100/16$, compared with $\sigma = 5$; the above approximation then gives 1.25, compared with the more exact value, 1.21.

The upper limit of $1 + m_0$ may be checked by reference to the distribution for N in the limiting case, as members of the same cluster may then all be assumed counted in a square whenever the centre falls within it. For Poisson families, this gives the probability-generating function of N,

$$\pi(z) = \exp\{\lambda A[e^{m_0(z-1)} - 1]\}$$ (11)

from which we find a mean $\lambda m_0 A$ and a variance of $\lambda m_0 A + \lambda m_0^2 A$, or a variance-mean ratio of $1 + m_0$, as above.

Finally, if the variance-mean ratio for a single square of side h is denoted by R_h, and the number of such squares is 2^{2k}, then an individual mean square in the χ^2 table has from its construction the form

$$(2^{2k}R_h - 2^{2(k-1)}R_{2h})/3.2^{2(k-1)}.$$ (12)

It should be noted that for these clustering-type models the mean square rises to a maximum when the spread of the clusters is small; in Simulation 1 even the smallest square unit gives a theoretical mean square which is 2.12, in contrast with Simulation 2, where square units with sides four times as long a theoretical mean square of 1.93, consistently with the five-fold scale increase in cluster size. In any empirical analysis a further increase in mean square could indicate a further clustering effect at a larger scale. To illustrate this type of analysis with real data, K. A. Kershaw's *Nardus Stricta* data (1957), discussed further by Mead (1974), have been taken. These were actually transect data of the cover type for a five point grid, and effectively one-dimensional, so that the nested analysis proceeds by factors of two. The data for the first transect are shown in Table VI (where the single row of counts is shown as eight consecutive rows for convenience). The clustering is evident from these counts, and the resulting mean squares rise to a high maximum of over 6, from the between 4's line onwards. While there is a still higher peak of 12, this is on the one transect based on only two degrees of freedom, and would not justify further consideration, except that it is maintained on the analysis (quoted from Mead) on the full data of five transects. This raises the question whether such a value can be precisely tested, once the original random hypothesis has been discarded.

A method due to Mead of handling this question may be illustrated on the data shown in Table VI. Let us tabulate the total counts of 4's viz

0,2,2,0; 0,0,1,10; 11,1,0,2; 5,9,4,10;

0,1,17,3; 0,3,0,0; 0,0,0,1; 0,0,0,5.

The differences between the neighbouring pair totals within sets of four are consequently 0, 11, 10, 0, 19, 3, 1, 5: total 49. If, however, the grouping of these pairs within sets of four is of no real consequence, a randomization within these sets would produce comparable totals. Each set of four produces three possible differences, and the eight sets 3^8 possible totals, in comparison with which the observed total of 49 may be judged. Mead also notes that the test may be modified if thought advisable, to reduce any effect of heterogeneity of scale within the groups of 4, by replacing the actual triads of possible differences, which are necessarily of the form (a, a, b), (a, b, c), or (a, b, b) where $a < b < c$, by $(0, 0, 2)$, $(0, 1, 2)$ and $(0, 2, 2)$ respectively.

With the results shown in Table VI the main interest with this further test is for the additional peak referred to above,

Table VI   K. A. Kershaw's *Nardus Stricta* data (first transect, adapted from R. Mead, Table 2 *Biometrics*, **30** (1974) 293)

					Total
(1)	· · · ·	· · · 2	2 · · ·	· · · ·	4
(2)	· · · ·	· · · ·	· · · 1	2 4 4 ·	11
(3)	2 3 4 2	1 · · ·	· · · ·	2 · · ·	14
(4)	· · 1 4	3 1 3 2	2 1 · 1	3 5 2 ·	28
(5)	· · · ·	· ° · 1	5 3 4 5	2 1 · ·	21
(6)	· · · ·	2 · 1 ·	· · · ·	· · · ·	3
(7)	· · · ·	· · · ·	· · · ·	· 1 ·[1] ·	1
(8)	· · · ·	· · · ·	· · · ·	1 · 2 2	5
					87

χ^2 *analysis*

	D.F.	S.S.	M.S.	Full data (cf. Mead)
Within 2's	64	56.60	0.88	0.99 (512)
Between 2's within 4's	32	39.36	1.23	2.38 (256)
Between 4's within 8's	16	87.72***	5.48	5.19 (128)
Between 8's within 16's	8	56.74***	7.09	4.72 (64)
Between 16's within 32's	4	26.90***	6.72	5.89 (32)
Between 32's within 64's	2	24.21***	12.11	13.41 (16)
Between 64's	1	8.39**	8.39	6.95 (8)
Total	127	299.92***	2.36	
M.S. excluding first two entries (from full data,			6.58 6.00)	

for which the full data must be used. The significance level obtained by Mead was 0.032 for the untransformed data, but only 0.092 for the distribution-free modified test, so that the significance of this peak must remain somewhat doubtful.

To avoid the complications of the exact randomization, Mead also gives the normal approximation in the distribution-free case.

Triads of type	(a, a, b),	(a, b, c),	(a, b, b)
Observed	n_1	n_2	n_3
	n_{10} a's	n_{20} a's	n_{30} a's
	n_{12} b's	n_{21} b's	n_{32} b's
		n_{22} c's	

For such a set of results, we should calculate

$$X = \frac{n_{21} + 2(n_{12} + n_{22} + n_{32}) - (2n_1 + 3n_2 + 4n_3)/3}{\{[4(n_1 + n_3) + 3n_2]/3\}^{1/2}}, \tag{13}$$

which is asymptotically normal with zero mean and unit variance.

3.2.1 Spectral analysis

For the spectral analysis of the (first) simulation referred to in §3.2, which was reported in the paper cited (Bartlett, 1964b), the 'periodogram' for such a point process realization of n points with co-ordinates x_s, y_s is calculated as

$$I_{pq} = J_{pq}J_{pq}^* = A_{pq}^2 + B_{pq}^2, \tag{14}$$

where

$$J_{pq} = A_p + iB_p = \sqrt{\frac{2}{\mu A}} \sum_{s=1}^{n} \exp\{i(x_s\omega_{1(p)} + y_s\omega_{2(q)}\}. \tag{15}$$

(ω_1, ω_2) being chosen to have a set of integral values $\omega_{1(p)} = 2\pi p/n$, $\omega_{2(q)} = 2\pi q/n$. To standardize the calculations, we may set $x_s' = nx_s/L_1$, $y_s' = ny_s/L_2$; this helps to eliminate bias near $\omega = 0$ due to the discrete component $2n^2/(\mu A)$ at the origin. In order that $E\{I_{pq}\} \to 2$ as $\omega \to \infty$, we choose also $\mu A = r$.

For processes with spectral density function, such as the two-dimensional Poisson process or a clustering process like (3) of §3.2, it is well-known (cf. Bartlett, 1966 §9.23 for one-dimensional point processes) that I_{pq} requires 'smoothing' if it is to converge to $g(\omega)$ as $n \to \infty$. Uniform weighting of individual values will usually be most convenient, and to avoid excessive tabulation of all I_{pq} values, cumulative totals of I_{pq} over blocks 6 x 6 in p and q were computed and are given in Table VIIa. The values of p and q chosen should in general cover the complete range including the axes $p = 0$ and $q = 0$ (but excluding $p = q = 0$) and *either p or q = $-1, -2, \ldots$*. However with this isotropic simulated case there seemed no point in including negative p or q values, which were taken in the range p, q up to 48.

The spectral values agree sufficiently well with the theoretical expected values, as is indicated in Table VIIb, where the values have been scaled by the expected values obtained by integration of the spectrum (3) of §3.2 to give a uniform expectation of 72 (2 x 36) equal to the asymptotic values for large p, q.

The use of spectral analysis seems theoretically more attractive than analyses on counts, which we have seen may be difficult to interpret, but of course requires precise pin-pointing of the data, and this is not always feasible. A further difficulty may be the computing involved, which gets more heavy in two-dimensional examples; it may therefore be useful to note an ingenious modification of the spectral analysis of point processes in one dimension proposed by French and Holden (1971) in the context of the analysis of neural impulses. In place of $dN(t)$ we introduce the continuous process

$$X(t) = \int \frac{\sin[(t - \tau)\pi]}{(t - \tau)\pi} \, dN(\tau), \tag{16}$$

which filters out the frequencies $\omega < \pi$, and then sample at

Table VIIa (From Bartlett, 1964b, Table 10)

Spectrum totals

$p\rightarrow$	186.45	182.91	110.88	76.51	64.85	51.72	79.27	81.37
q	111.37	130.41	98.34	67.57	42.39	55.63	58.04	59.62
$\downarrow$	109.14	85.28	74.79	61.97	49.43	66.43	66.37	68.33
	81.84	65.22	52.60	76.49	49.31	70.44	92.37	65.02
	66.30	62.62	63.10	67.12	63.62	59.16	62.35	66.53
	80.26	88.83	90.23	92.42	74.18	89.91	87.16	91.75
	76.25	72.46	82.16	65.90	79.26	87.36	72.43	108.92
	63.08	92.32	83.08	55.71	77.24	60.52	78.10	79.95

Table VIIb (From Bartlett, 1964b, Table 11)

Spectrum totals scaled to give expectation of 72

$p\rightarrow$	67.19	79.01	63.91	58.18	58.95	50.46	78.88	81.37
q	48.11	65.86	63.65	54.93	39.43	54.54	57.75	59.62
$\downarrow$	62.90	55.20	57.31	54.84	43.74	65.77	66.37	68.33
	62.24	53.02	46.55	72.50	46.74	70.09	92.37	65.02
	60.27	58.25	60.67	65.80	63.30	59.16	62.35	66.53
	78.30	87.09	89.34	91.96	74.18	89.91	87.16	91.75
	75.87	72.10	82.16	65.90	79.26	87.36	72.43	108.92
	63.08	92.32	83.08	55.71	77.24	60.52	78.10	79.95

unit time intervals, which permits the use of fast Fourier transform computing techniques (Cooley and Tukey, 1965). Any aliasing problem for frequencies over π is avoided, as they have already been eliminated. The frequency cut-off may be chosen by appropriate choice of the time unit.

This device is obviously applicable in more than one dimension. Thus for two spatial dimensions we should write

$$X(\mathbf{r}) = \int \frac{\sin[(x_1 - x_2)\pi]}{(x_1 - x_2)\pi} \; \frac{\sin[y_1 - y_2)\pi]}{(y_1 - y_2)\pi} \; dN(\mathbf{s}) \qquad (17)$$

where $\mathbf{r} = (x_1, y_1)$, $\mathbf{s} = (x_2, y_2)$, and sample $X(\mathbf{r})$ on the intersections of the rectangular lattice of unit squares.

3.2.2 Nearest-neighbour distances between gulls' nests

The general effect of inhibitory action between individuals on spectral analysis or related analysis of variance results has

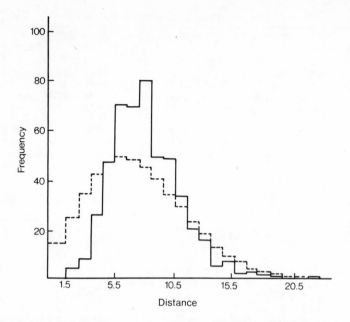

Fig. 9 *Nearest-neighbour nesting distances for the lesser black-backed gull. The mean distance is 8.069, and the dotted histogram is a random distribution with the same mean (From Adv. Appl. Prob., 1974a)*

been already mentioned at the end of § 1.2.1. Sharply defined nearest-neighbour inhibition is more directly studied by measurements between neighbours, but the relevant theoretical models to be fitted to such measurements usually have to be studied by simulation (cf. § 1.2.2). Two-dimensional examples of such data are the positions of gulls' nests, or the positions of towns.

In Table VIII and Figs. 9 and 10 are shown the observed frequency distributions of nearest-neighbour radial distances (r) from each nest for two gull species (a) the lesser black-backed gull (b) the herring gull.

In Figs. 9 and 10 the theoretical *random* distributions which are well-known to have density function

$$f(r) = 2\pi\lambda r\, e^{-\pi\lambda r^2} \tag{18}$$

are also shown (for boundary values of r 1.5, 2.5, . . .), and

Table VIII (MacRoberts' data)

r	(a)	(b)	r	(a)	(b)
1	0	0	14	5	11
2	5	7	15	7	7
3	8	4	16	2	20
4	27	8	17	3	11
5	48	28	18	2	10
6	71	34	19	1	3
7	70	39	20	0	5
8	79	68	21	0	8
9	48	50	22	1	2
10	47	55	23		1
11	33	51	24		1
12	20	41	25		1
13	16	33	26		1
			Total	493	499

are obviously unsatisfactory for small r, owing to the tendency for very small distances to be avoided. If the Eberhardt Index (Eberhardt, 1967)

$$n\Sigma r^2 /(\Sigma r)^2$$

is calculated, it gives values 1.133 for (a), and 1.161 for (b) compared with the theoretical value 1.27 for a random distribution. The 95% confidence interval for a number of simulations for $n = 500$ carried out by P. D. Macdonald was 1.23–1.33 (note that even in the random case the known nearest-neighbour distance distribution does not give us the *simultaneous* distribution determining the standard error).

An attempt was made to allow for local inhibition in a further simulation with $n = 500$. Unfortunately, the total areas involved were not known, and once an inhibition effect is included the scale is of course fixed. A further difficulty is the indication from the data that any cut-off at small distances is not sharp; and in case (a) the model adopted had

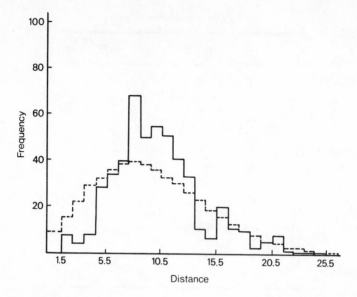

Fig. 10 *Nearest-neighbour nesting distances for the herring gull. The mean distance is 10.2485, and the dotted histogram is a random distribution with the same mean (From Adv. Appl. Prob., 1974a)*

the variable inhibition radius

r:	0	1	2	3	4
p	0	0.10	0.10	0.10	0.70

The resulting distribution from simulation ($n = 500$) gave of course reasonable agreement for very small r (as the variable inhibition radius had been arrived at after inspection of the data), but too large a mean r (10.553), and hence too long a tail. A rough adjustment was made by re-scaling the distances (using equivalent areas, but with no additional allowance for non-uniform spread) to make the means agree; the small r values are now inappropriate (in the first place because of the shifted inhibition effect, but also because of the crude nature of the re-scaling), but the fit at the other end seems now reasonable (see Fig. 11 and Table IX), so that in spite of these difficulties of comparing with an inappropriate simulation, the inhibition effect seems to emerge as the main cause of the non-random distribution.

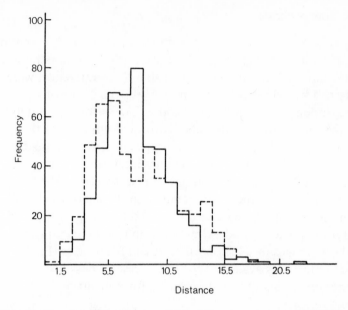

Fig. 11 *Simulated distribution (dotted histogram) compared with observed distribution of nearest-neighbour distances after re-scaling to give closer agreement for the mean (which is now 8.007 for the re-scaled simulation). (From Adv. Appl. Prob., 1974a)*

Table IX (From Bartlett, 1974a, Table VII) Simulation for (a)

r	Simulated	Re-scaled	r	Simulated	Re-scaled
1	0	2	14	21	26
2	4	9	15	29	13
3	9	20	16	11	7
4	15	49	17	17	3
5	39	66	18	14	1
6	37	67	19	9	
7	61	45	20	7	
8	48	34	21	11	
9	36	48	22	6	
10	30	35	23	3	
11	32	33	24	2	
12	29	22	25		
13	31	20	26		
			Total	500	500

3.3 Line processes

The spectral analysis of the p, θ representation of the random set of 50 lines of Fig. 2 (§1.3) is given below for illustration. These data, which are quoted in Table X for reference, were obtained by calculating $\theta = \tan^{-1}(x/y)$ from a pair of independent normal variates x and y (Tracts for Computers, No. 25) and p from uniformly distributed numbers in the range 0 to 100 (Tracts for Computers, No. 24), dividing by $\sqrt{2}$ for Fig. 2 to ensure that the lines intersected a circle of radius $50\sqrt{2}$. (While two did not intersect the inscribed square of side 50, they were retained in the analysis.) It will be noticed that the representation used here is for θ 0 to 2π, and p the range 0 to 50 extracted from 0 to ∞, in contrast with the ranges in §1.3 of θ 0 to π, and p $-\infty$ to ∞. Either representation is available, but in the case of possibly non-random real data the $-\infty$ to ∞ range for p might be preferable.

The periodogram sums are analogous to those defined in §3.2.1, modified to cope with the variable θ by defining (in the case of the p, θ representation)

$$J_{qs} = \frac{2}{n} \sum_{r=1}^{n} \exp\{i(x_r \omega_q + \theta_r s\}, \qquad (19)$$

where $\omega_q = 2\pi q/n$, $s = 0, \pm 1, \pm 2, \ldots$ The theoretical spectral

Table X (From Bartlett, 1967b, Table I)

p	θ	p	θ	p	θ	p	θ	p	θ
2.45	2.461	48.32	4.683	12.40	0.913	5.60	0.217	28.76	5.931
11.85	5.771	19.39	2.244	29.15	1.154	28.74	3.630	36.19	2.799
39.31	4.584	21.17	1.534	64.52	1.856	66.72	3.074	41.70	3.879
59.69	4.893	7.97	1.637	60.26	0.392	12.99	4.820	7.69	5.280
23.48	0.017	27.03	4.223	62.76	0.484	10.63	0.582	43.59	3.436
12.86	0.557	47.14	0.000	30.92	3.261	21.31	1.443	54.86	4.962
16.56	3.737	48.12	0.883	44.03	4.573	26.77	0.671	37.16	0.173
26.76	5.110	45.39	4.009	39.85	5.085	69.96	5.808	26.91	0.375
40.04	0.983	5.64	1.540	12.40	1.346	67.00	5.945	3.10	2.922
11.37	5.1–2	19.08	3.038	8.73	0.116	11.92	1.307	56.70	5.731

Table XI

$q = 1\text{--}10$	$11\text{--}20$	$21\text{--}30$	$31\text{--}40$	$41\text{--}50$	$51\text{--}60$	$61\text{--}70$	$71\text{--}80$	$81\text{--}90$	$91\text{--}100$
$s = -5$ 14.20	22.97	24.27	18.36	18.80	34.06	26.75	14.57	22.21	14.80
-4 20.19	16.36	17.79	23.92	19.70	20.39	18.63	15.00	21.84	29.55
-3 24.81	23.18	16.57	22.39	19.83	16.33	11.76	13.96	30.37	18.02
-2 13.84	20.37	19.65	25.24	22.12	8.99	19.38	18.21	13.76	14.24
-1 23.31	13.33	25.54	15.48	26.40	20.69	18.50	11.60	16.15	25.08
0 24.86	20.14	29.06	39.67	13.77	10.39	27.67	22.12	20.41	8.60
1 29.30	9.32	11.85	31.77	15.19	11.34	16.53	25.78	17.37	15.19
2 19.81	23.92	16.38	21.54	19.12	27.23	13.16	17.83	26.73	12.29
3 15.27	37.48	29.85	13.96	27.14	32.64	25.34	18.36	27.30	40.61
4 11.25	12.67	30.97	14.65	22.84	21.22	27.40	10.73	17.71	41.86
5 19.06	30.64	23.18	16.94	12.93	13.69	23.81	22.92	16.66	23.55

function is of the form

$$\alpha_s(\omega) = 1 + f(\omega)\delta_s \tag{20}$$

for random independent θ, where for $\theta = 0$ to $2\pi\delta_s = 0$ except for $s = 0$, when $\delta_s = 1$. Here $f(\omega)$ is the modulated part of the spectral function of p, and is 0 for random p ($\alpha_s(\omega)$ being then standardized to unity).

Values of $I_{qs} = J_{qs}J^*_{qs}$ are shown in Table XI in block totals of ten for $q = 1$ to 100 and $s = 0$ to ± 5, and appear satisfactorily uniform. The mean value of these totals is 21.0, or 2.10 for each I_{qs}, compared with the expected value of 2.00.

CHAPTER FOUR

Analyses of Processes on a Lattice

4.1 Lattice processes X_i. Continuous variables. Examples using simultaneous and conditional models

In Whittle's pioneering 1954 paper, he discussed the numerical fitting of models like (5) of §2.1, viz

$$X_{rs} = \beta_1 (X_{r-1,s} + X_{r+1,s}) + \beta_2 (X_{r,s-1} + X_{r,s+1}) + Y_{rs},$$

(1)

where the Y_{rs} are assumed independent (and with zero mean), to the classical Mercer and Hall uniformity trial on wheat. He noted the limitations of this set of data, which were treated as values X_{rs} when they are really of the integrated type (see equation (11) of §1.1).

Other examples include studies by Mead (e.g. 1967) in the context of inter-plant competition, though there seems, as already remarked (see §2.1), to have been some confusion in the literature between model (1) and the conditional nearest-neighbour model

$$E\{X_{rs} \mid \text{all other } X\text{'s}\} = \beta_1 (X_{r-1,s} + X_{r+1,s})$$
$$+ \beta_2 (X_{r,s-1} + X_{r,s+1}) \qquad (2)$$

which was first fitted to some of these examples by Besag (1974a).

The Mercer and Hall data. These data consist of grain yields from 500 plots, 11 ft by 10.82 ft, arranged in a 20 x 25 rectangle. Estimated autocorrelations for these data as given by Whittle are quoted in Table XII, the co-ordinate s

Table XII (From Whittle, 1954, Table 1)

t	$s = 0$	$s = 1$	$s = 2$	$s = 3$	$s = 4$
−3	0.1880	0.1602	0.1509	0.1276	0.1352
−2	0.1510	0.0234	0.0020	−0.0137	−0.1039
−1	0.2923	0.1853	0.1349	0.0788	0.0878
0	1.0000	0.5252	0.4055	0.3639	0.3561
1	0.2923	0.2354	0.1799	0.1205	0.1399
2	0.1510	0.1285	0.0999	0.0749	0.0859
3	0.1880	0.1935	0.2483	0.2415	0.2284

denoting the north-south direction and t the east-west. In the case of fitting the simultaneous model (1), Whittle showed that (for normal processes) the maximum likelihood estimation of the coefficients is equivalent to the minimization of the usual mean squared residual U times a function k of the coefficients, this last factor arising from the Jacobian of the transformation from Y_{rs} to X_{rs} (see §2.1). In the case of one-sided schemes, k is of course unity. For models with spectrum

$$[1 - H(z_1, z_2)]^{-2}, \tag{3}$$

it was shown that $\log k$ is the absolute term in the expansion of

$$-2 \log[1 - H(z_1, z_2)]. \tag{4}$$

For the model (1), we have

$$\log k = \sum_{i=1}^{\infty} \sum_{j=0}^{i} \frac{(2i)!}{i[j!(i-j)!]^2} \beta_1^{2i} \beta_2^{2j}, \tag{5}$$

which in the completely symmetric case $\beta_1 = \beta_2 = \beta$, say reduces to

$$\sum_{i=1}^{\infty} \frac{1}{i} \left(\frac{2i}{i} \right)^2 (\beta)^{2i}. \tag{6}$$

In the case of conditional models, Besag (1974a) has noted that Whittle's method of obtaining maximum likelihood estimates may still be applied. Thus for a process with

Table XIII

Description of model	Whittle's model no.	β_1	β_2	k	U	kU
One-sided	1	0.488	0.202	1	0.6848	0.6848
Completely symmetric	5	0.159	0.159	1.1240	0.6508	0.7314
Two-sided	6	0.213	0.102	1.1332	0.6217	0.7045

spectrum as in (3), but raised to the negative first power only, the appropriate k is merely half the value in (4). The maximum likelihood method in this case may be shown from the functional form of the auto-normal scheme to equate the relevant theoretical and observed autocorrelations.

Some of Whittle's original analyses are summarized in Table XIII. The one-sided model referred to was

$$X_{rs} = \beta_1 X_{r+1,s} + \beta_2 X_{r,s+1} + Y_{rs}. \tag{7}$$

The interesting feature of these three first-order models is that the one-sided model gives the best fit; and it has been pointed out by Besag that this might be construed as indicating a better fit from the conditional (or auto-normal) first-order model, in the sense that the one-sided model was shown in §2.1 to provide a first approximation to the conditional two-sided model. The corresponding approximate estimates of β_1 and β_2 for the two-sided model are shown in Table XIV, which summarizes the various estimates obtained by Besag for the auto-normal model, including the second

Table XIV Auto-normal model

	β_1	β_2
Maximum likelihood	0.368	0.107
Coding: (1)	0.332*	0.128*
(2)	0.354*	0.166*
Mean of (1) and (2)	0.343	0.147
1st one-sided approximation	0.382	0.158
2nd one-sided approximation	0.374	0.119

*s.e. 0.03

approximation via the one-sided model, the estimates using the coding methods described in §2.1 and the maximum likelihood estimates. These estimates appear reasonably consistent with each other.

Second-order models　The values of kU in Table XIII do not include the smallest among Whittle's models, as he considered adding further terms, and in particular with the one-sided model (his Model 4)

$$X_{rs} = \beta_1 X_{r+1,s} + \beta_2 X_{r,s+1} + \gamma_1 X_{r+2,s} + \gamma_2 X_{r,s+2} \qquad (8)$$

obtained estimates

$$\beta_1 = 0.402, \beta_2 = 0.168, \gamma_1 = 0.172, \gamma_2 = 0.092$$

with a value of $kU = 0.6564$, $(k = 1)$.

Besag also fitted an auto-normal model with further terms relating to diagonal nearest-neighbours viz.

$$\begin{aligned}
E\{X_{rs} \mid \text{all other values}\} &= \beta_1 (X_{r-1,s} + X_{r+1,s}) \\
&+ \beta_2 (X_{r,s-1} + X_{r,s+1}) \\
&+ \gamma_1 (X_{r-1,s-1} + X_{r+1,s+1}) \\
&+ \gamma_2 (X_{r-1,s+1} + X_{r+1,s-1})
\end{aligned}$$
$$(9)$$

using the second coding pattern given in Fig. 6 (§2.1) as appropriate for this model. Four possible sets of patterns give rise to four sets of estimates, which are quoted in Table XV,

Table XV (From Besag, 1974a, Table 10)

Coding pattern	β_1	β_2	γ_1	γ_2
(i)	0.344	0.043	0.079	−0.062
(ii)	0.318	0.085	0.016	0.011
(iii)	0.407	0.243	−0.067	−0.034
(iv)	0.361	0.236	−0.092	−0.041
s.e.	0.05	0.06	0.07	0.06
Mean estimate	0.358	0.152	−0.016	−0.032

Table XVI (From Besag, 1974a, Table II) Analysis of
variance (pattern (i))

	Sum of squares	D.F.	Mean Square
β_1, β_2	9.63	2	4.81
γ_1, γ_2	0.19	2	0.10
Residual	10.89	103	0.106
Total	20.71	107	

together with the s.e.'s of each set. It appears that there is no
significant gain by fitting the additional γ-coefficients in (9),
and this is indicated also by the analysis of variance of the
first set (Table XVI). The coefficients in Whittle's model (8)
have of course no very close relation with the four
coefficients in Besag's model (9).

Both Whittle and Besag draw attention to trends in the
data needing further consideration before any final
conclusions on the best-filling model are drawn. It should be
noticed that the values of β_2 in Table XI appear inconsistent
when the (iii) and (iv) sets are compared with (i) and (ii).

4.2 Discrete variables. Simple χ^2 analyses. Further analyses and examples.

Coming next to lattice processes involving discrete variables,
which will often, but not always, be binary, I shall first of all
in this section note some of my first χ^2 analyses of such data.
These analyses, whilst the further developments outlined in
Part I have enabled us to see their rôle and limitations in a
wider theoretical context, have the merit of being both
informative and simple, so that they are likely to be still
useful even for situations where there may be more logical,
but, from the analysis viewpoint at least, more complicated,
models.

Like the continuous variable example of the Mercer and
Hall data, these first two examples to be considered are not
strictly lattice data, the lattice being imposed on a continuous

area, and, with the present examples, counts taken over each cell of the lattice constituting the data. Such data may perhaps more adequately be analysed by simple semi-empirical models with a view to studying their structure than by a more complicated model analytically only strictly appropriate to genuine lattice data.

The two sets of data are similar in type, the first being numbers of balsam-fir seedlings in five feet square quadrats taken from Ghent (1963), and the second counts of *Carex arenaria* made by Dr. P. Greig-Smith and quoted in Bartlett (1971a). The data are thus not dissimilar in type from the simulated clustering model of §3.2, but the χ^2 analyses illustrated there were mainly checks on complete randomness, whereas here it is the use of models of local relations that necessitates the care on the interpretation and validity of the χ^2 analysis.

It will be seen that in Table XVII the counts range up to 7, and the binary presence-absence classification would be throwing away rather too much information in this case. Regression on the actual neighbouring counts might be a possibility, but as simple linear nearest-neighbour models are to be tried, a reasonable compromise was to follow Ghent in classifying the counts into *three* categories, L, low density (0 or 1), M, medium (2 or 3), and H, high (4 or more). We

Table XVII Number of balsam-fir seedlings in five feet square quadrats (Taken from Ghent, 1963 Fig. 4)

0	1	2	3	4	3	4	2	2	1
0	2	0	2	4	2	3	3	4	2
1	1	1	1	4	1	5	2	2	3
4	1	2	5	2	0	3	2	1	1
3	1	4	3	1	0	0	2	7	0
4	2	0	0	2	0	3	2	3	2
2	2	2	0	3	4	7	4	3	3
2	3	1	2	3	8	5	5	1	2
1	1	2	1	4	4	5	3	2	3
3	1	6	1	3	5	4	7	4	3

Table XVIII (From Bartlett, 1967a, Table 9)

$X = -1$	2(4.00)	11 (9.82)	8 (6.58)	3(3.04)	0(0.58)	24
0	6(3.55)	7(11.10)	13(10.21)	7(7.55)	3(3.55)	36
1	0(0.45)	7 (4.07)	2 (6.21)	7(6.41)	5(3.87)	21
Total	8	25	23	17	8	81

then, if we classify the counts in relation to the *one-sided* pattern of nearest-neighbours (*L, M* or *H*) to the left and above in the Table, and consider moreover an assumed dependence of X_{rs} on the sum $S_{rs} = X_{r-1,s} + X_{r,s-1}$, scoring *L* as -1, *M* as 0, and *H* as 1, arrive at the frequency classification (for the last nine rows and columns) in Table XVIII. A χ^2 analysis on the hypothesis of no relation with *S*, so that the row totals are used to provide the expected frequencies for each column, gives 19.68 with 8 d.f. ($P \sim 0.01$). However, if the probability model

$$P\{X_{rs} = r \mid S_{rs} = s\} = \alpha_r + \beta_{rs} \tag{10}$$

is fitted, the coefficient β being taken as the 'unweighted' regression estimate 0.107, we arrive at the expected frequencies shown in brackets in Table XVIII. χ^2 is now 11.94 with 7 d.f. ($P \sim 0.10$). The fit could perhaps be improved further by the use of a more efficient estimate of β or by the use of a less arbitrary (though admissible) model, but seems sufficiently satisfactory.

With Greig-Smith's data (Table XIX), the counts are small enough for presence-absence analysis to seem reasonable. A frequency table of the 0, 1 values classified against the one-sided nearest-neighbour configurations gave the results of Table XX. In this Table the expected frequencies in brackets are based on the simple one-sided model

$$P\{x_{rs} = 1 \mid x_{r-1,s}, x_{r,s-1}\} = p + \alpha(x_{r-1,s} + x_{r,s-1}) \tag{11}$$

with $\alpha = 0.1411$; and, while we know that this model cannot be an exact conditional nearest-neighbour model in the two-sided sense, it gives an adequate fit as far as Table XX is

Table XIX Counts of *Carex arenaria* (Newborough Warren, 1953—Dr. Greig-Smith) 24 × 24 grid of contiguous squares 10 cm × 10 cm.

Table XX (From Bartlett, 1971a, Table VI)

	0 0 .	1 0 .	0 1 .	1 1 .	Total
0	194(190.8)	55(60.7)	54(56.8)	36(31.4)	339
1	50 (53.2)	38(32.3)	33(30.2)	24(28.6)	145
Total	244	93	87	60	484*

*A 22 x 22 array, the last row and column being omitted for symmetry with the *given* first row and column.

	39	40	41	42	43	44	45	46	47	48	49
1	•	•	•	•	•	N	V	N	•	N	N
2	•	N	•	•	N	N	N	•	N	•	•
3	N	N	N	N	•	N	•	N	N	N	•
4	N	N	•	N	N	N	N	N	N	•	•
5	N	N	N	N	•	N	•	•	N	N	•
6	•	N	N	•	N	N	N	•	•	N	N
7	N	N	N	N	•	N	•	•	•	•	•
8	N	•	V	N	•	N	N	N	N	N	N
9	N	N	•	N	V	•	N	N	•	N	•
10	•	N	N	N	N	N	N	N	V	•	N
11	N	N	N	•	•	•	N	N	•	•	N
12	•	•	N	•	•	N	•	N	N	N	N
13	N	N	V	N	•	N	•	•	N	•	N
14	N	•	N	N	•	N	V	•	V	•	•
15	•	V	•	•	•	•	•	•	•	N	•
16	•	•	•	•	N	•	V	•	N	N	N
17	•	•	•	•	•	•	•	•	•	V	N
18	•	∘	N	∘	N	•	•	•	•	N	•
19	N	•	•	N	•	•	•	•	•	•	•
20	∘	∘	∘	N	∘	∘	∘	∘	N	∘	•
21	∘	∘	V	∘	∘	N	∘	•	N	•	•
22	•	∘	∘	∘	∘	N	•	N	N	∘	N
23	∘	∘	∘	∘	N	V	•	•	∘	•	N
24	∘	∘	N	•	•	•	∘	•	∘	•	N
25	∘	•	∘	∘	∘	•	∘	N	∘	N	N
26	∘	•	•	∘	∘	∘	∘	∘	N	N	N
27	∘	•	•	•	∘	∘	∘	•	•	N	N
28	N	•	•	∘	∘	•	∘	•	•	N	∘
29	∘	∘	•	∘	•	∘	∘	∘	∘	N	∘
30	•	•	∘	∘	∘	•	∘	∘	∘	∘	∘

N = Nettlehead plant.   V = Vacancy.   • = Plant without symptoms.

Fig. 12 *Incidence of nettlehead virus in 1951 among hop plants. (From Freeman, 1953, Biometrika, Fig. 1)*

Table XXI (From Bartlett, 1974a, Table VIII) χ^2 *analyses of Freeman's data* 1951 plants (Lattice 11 x 30, with effect of 12 vacancies neglected)

		One-sided nearest-neighbour configurations				
		0 0	1 0	0 1	1 1	
First	0	13(10.5)	17(16.9)	15(17.8)	19(18.7)	64
half	1	10(12.5)	20(20.1)	24(21.2)	22(22.3)	76
		23	37	39	41	140

Hypothesis of no dependence: $p_1 = 0.5429$ $\chi_1^2 = 1.92$ (3 d.f.)

Second		83(77.5)	14(18.6)	12(9.7)	4(7.3)	113
half		13(18.5)	9 (4.4)	0(2.3)	5(1.7)	27
		96	23	12	9	140

Hypothesis of no dependence: $p_2 = 0.1929$ $\chi_2^2 = 18.73$*** (3 d.f.)

Total	0	96(75.2)	31(37.9)	27(32.2)	23(31.6)	177
	1	23(43.8)	29(22.1)	24(18.8)	27(18.4)	103
		119	60	51	50	280

Hypothesis of no dependence: $p = 0.3679$ $\chi^2 = 27.66$*** (3 d.f.)

concerned ($\chi^2 = 3.60$ with 2 d.f.). This does not imply of course that a more detailed look at the data would not reveal discrepancies (cf. the analysis below of Gleaves' data).

 The third example in this section is a more genuine case of lattice data, consisting of records in 1951 of the presence or absence of nettlehead virus disease of a lattice of hop plants. There are, however, as with most real data, other complications militating against too precise an analysis, including some vacancies where there are no plants (v in the diagram, Fig. 12), and an obvious decrease in incidence from top right to bottom left. This would make any overall tests of 'runs' or 'links' to demonstrate non-randomness very misleading. The χ^2 analysis which is shown in Table XXI and based on a one-sided nearest-neighbour classification has been

Table XXII (From Bartlett, 1974a Table IX) χ^2 analysis of regression model

Regression of 1951 status (if not diseased in 1950) on 1950 (x is total number of diseased nearest-neighbours, including diagonals)

x:	0	1	2	3	4	5	6	
Diseased	14(28.7) (14.7)	9(12.4) (10.8)	15 (9.1) (11.1)	11 (7.8) (12.3)	11(4.9) (9.6)	4(2.3) (5.4)	2(0.8) (2.1)	66
Total	111	48	35	30	19	9	3	255

No dependence (upper expectations, last two groups pooled)

$\chi_1^2 = 32.24^{***}$ (5 d.f.)

Regression model (lower expectations)

$p = 0.2588 + 0.0931 (x - 1.361)$

$\chi_2^2 = 3.88$ (5 d.f.).

Table XXIII (From Besag, 1974a,
Table 1)

Coding pattern	α	β
(i)	−2.254	0.724
(ii)	−2.141	0.748
Approx. s.e.	0.07	0.04
Mean for (i) & (ii)	−2.198	0.736

carried out separately for the top and bottom halves of the data, and the significance of the χ^2 for the entire data is attributable partly to the effect of pooling heterogeneity and partly to the evidence of association *where incidence is low*.

The citing of this example here is partly to illustrate (Table XXII) a fit of a temporal-spatial model conceived in the spirit of the temporal models referred to in §2.1. The existence of records for 1950 enabled a regression to be made on nearest-neighbours the year before. This eliminated the complication in the χ^2 analysis if symmetrical nearest-neighbours are to be tested; it also enabled diagonal nearest-neighbours to be readily included, at first separately, but finally (and adequately) on a par with the others.

It will be seen that the fit of the regression model (in spite of its over-simple *linear* form) appears adequate.

If a symmetrical spatial model is to be tested on data of this type, the difficulty of extracting independent degrees of freedom (see Bartlett, 1971a) may be avoided, at the cost of some inefficiency, by restricting the analysis to alternate sites, a method due to Besag (see §2.1), who first examined the fit of the auto-logistic model to the data of Greig-Smith's used as one of my earlier examples.

Besag's auto-logistic analysis of Gleaves' data. Some data similar in type to Greig-Smith's data were collected by Gleaves on *Plantago lanceolata*. The area sampled at Treloggan, Flintshire was a long and narrow transect 10 x 940 of cells 2 x 2 cm, presence/absence of plants in each cell being the variable considered by Besag (1974a) in his analysis,

Table XXIV (From Besag, 1974a, Table2)

x_{rs}	Frequencies $z_{rs} = 0$	Coding pattern (i) (out of 2)				Total
		1	2	3	4	
0	1798(1786.5)	847(856.6)	329(338.7)	101(98.9)	25(19.3)	3100
1	176 (187.5)	195(185.4)	161(151.3)	89(91.1)	31(36.7)	652
Total	1974	1042	490	190	56	3752

$\chi^2 = 4.84$ (2 d.f.)

Table XXV (From Besag 1974a, Table 4)

Coding pattern	α	β	γ
(i)	−2.534	0.518	0.497
(ii)	−2.459	0.529	0.528
(iii)	−2.369	0.543	0.456
(iv)	−2.456	0.515	0.487
Approx. s.e.	0.10	0.07	0.07
Mean of (i) . . . (iv)	−2.455	0.526	0.492

which was based firstly on the isotropic auto-logistic model
(see §2.2)

$$P\{x_{rs} \mid \text{all other values}\} = \frac{\exp\{(\alpha + \beta z_{rs})x_{rs}\}}{1 + \exp(\alpha + \beta z_{rs})}, \qquad (12)$$

where $z_{rs} = x_{r-1,s} + x_{r+1,s} + x_{r,s-1} + x_{r,s+1}$. Estimates for α
and β were made for the coding pattern given §2.1 and are
quoted in Table XXIII. Table XXIV gives the observed and
expected frequencies for the first of the two analyses (the
second gives a similar result) and suggests a satisfactory fit,
though Besag notes that with only 2 d.f. available for the
goodness-of-fit test the class of alternatives under scrutiny is

Table XXVI (From Besag, 1974a, Table 6)

	Frequencies		Coding pattern (i) (out of 4)				
u_{rs}	x_{rs}	$z_{rs} = 0$	1	2	3	4
0	0	628(617.1)	207(220.6)	61(60.5)	6 (6.5)	1(1.8)
	1	38 (48.9)	43 (29.4)	13(13.5)	3 (2.5)	2(1.2)
1	0	223(218.5)	135(141.1)	61(56.3)	16(14.8)	3(3.4)
	1	24 (28.5)	37 (30.9)	16(20.7)	8 (9.2)	4(3.6)
2	0	49 (54.4)	60 (54.4)	36(37.4)	18(14.4)	4(3.3)
	1	17 (11.6)	14 (19.6)	24(22.6)	11(14.6)	5(5.7)
3	0	8 (8.9)	7 (6.9)	17(19.6)	7 (9.4)	4(1.8)
	1	4 (3.1)	4 (4.1)	22(19.4)	18(15.6)	3(5.2)
4	0	0 (0.0)	0 (0.0)	4 (3.8)	1 (0.5)	0(0.5)
	1	0 (0.0)	0 (0.0)	6 (6.2)	1 (1.5)	3(2.5)

necessarily very limited (cf. the one-sided analysis of Greig-Smith's data). The higher-order isotropic model

$$P\{x_{rs} \mid \text{all other values}\} \propto \exp\{\alpha + \beta z_{rs} + \gamma u_{rs}\} x_{rs} \quad (13)$$

where $u_{rs} = x_{r-1,s-1} + x_{r+1,s+1} + x_{r-1,s+1} + x_{r+1,s-1}$, was also considered by Besag, leading to the estimates of Table XXV using the four coding patterns available for this higher-order model. The frequency table for the first of these analyses is shown in Table XXVI, and gives a χ^2 of 28.3 with 20 d.f. (approximately, owing to some small expectations). Similar tests for all four analyses suggests a barely adequate fit, although it is clear from Table XXV that the inclusion of the γ coefficient has been useful.

It will be recalled that for the auto-normal process the maximum likelihood method is equivalent to equating the relevant observed and theoretical auto-correlations. This is readily shown to be true also for the auto-logistic model, but stresses the present unavailability of the maximum likelihood method for this model when the theoretical auto-correlations are still unknown for general α, β.

Appendix

Table AI Correlations $\overset{\circ}{\rho}_{rs}(2\beta)$ for the auto-normal model (symmetric case)

β	$\overset{\circ}{\rho}_{10}(2\beta)$	$\overset{\circ}{\rho}_{11}(2\beta)$	$\overset{\circ}{\rho}_{20}(2\beta)$
0.0200	0.0200	0.0008	0.0004
0.0400	0.0403	0.0032	0.0016
0.0600	0.0611	0.0074	0.0038
0.0800	0.0827	0.0135	0.0069
0.1000	0.1055	0.0218	0.0114
0.1200	0.1299	0.0327	0.0174
0.1400	0.1567	0.0470	0.0256
0.1600	0.1869	0.0656	0.0370
0.1800	0.2222	0.0907	0.0531
0.2000	0.2659	0.1260	0.0777
0.2100	0.2930	0.1500	0.0955
0.2200	0.3260	0.1810	0.1198
0.2300	0.3690	0.2242	0.1559
0.2400	0.4341	0.2946	0.2196
0.2410	0.4432	0.3048	0.2293
0.2420	0.4530	0.3160	0.2400
0.2430	0.4639	0.3284	0.2521
0.2440	0.4760	0.3425	0.2660
0.2450	0.4898	0.3586	0.2821
0.2460	0.5060	0.3777	0.3015
0.2470	0.5256	0.4011	0.3257
0.2480	0.5512	0.4321	0.3583
0.2490	0.5898	0.4795	0.4094
0.2495	0.6228	0.5207	0.4548
0.2496	0.6324	0.5328	0.4682
0.2497	0.6442	0.5475	0.4847
0.2498	0.6596	0.5669	0.5065
0.2499	0.6831	0.5967	0.5401

Table AI   (*continued*)

β	$\overset{\circ}{\rho}_{10}(2\beta)$	$\overset{\circ}{\rho}_{11}(2\beta)$	$\overset{\circ}{\rho}_{20}(2\beta)$
0.2499600	0.7098	0.6306	0.5784
0.2499875	0.7374	0.6660	0.6177
0.2499920	0.7469	0.6781	0.6315
0.2499980	0.7725	0.7106	0.6687
0.2499998750	0.8098	0.7580	0.7230
0.2499999550	0.8216	0.7731	0.7402
0.2499999950	0.8414	0.7982	0.7691
0.2499999992	0.8548	0.8153	0.7886
0.25−	0.9205	0.8989	0.8843
0.25−	0.9603	0.9494	0.9421
0.25−	0.9801	0.9747	0.9711
0.25	1.0000	1.0000	1.0000

Table AII Product moment ρ_{10} and m^2 for the auto-logistic model
(symmetric case with $\alpha = 0$, $x = \pm 1$)

$\epsilon' = -\tanh 2\gamma$	$\rho_{10}(\epsilon')$	ϵ'	ρ_{10}	m^2
0.7071	0.7071	0.7071	0.7071	0
0.7000	0.6661	0.7141	0.7474	0.5258
0.6971	0.6534	0.7170	0.7594	0.5714
0.6948	0.6445	0.7192	0.7720	0.5992
0.6911	0.6306	0.7228	0.7806	0.6368
0.6844	0.6086	0.7291	0.8003	0.6876
0.6747	0.5788	0.7381	0.8258	0.7412
0.6610	0.5458	0.7504	0.8518	0.7943
0.6502	0.5217	0.7598	0.8697	0.8253
0.6411	0.5030	0.7675	0.8827	0.8464
0.6329	0.4878	0.7743	0.8928	0.8627
0.6000	0.4332	0.8000	0.9251	0.9093
0.5514	0.3687	0.8343	0.9549	0.9485
0.5124	0.3259	0.8588	0.9700	0.9667
0.4782	0.2930	0.8782	0.9792	0.9772
0.4472	0.2659	0.8944	0.9850	0.9840
0.3162	0.1713	0.9847	0.9970	0.9969
0.2042	0.1053	0.9790	0.9995	0.9995
0.1005	0.0507	0.9949	1.0000	1.0000
0.0000	0.0000	1.0000	1.0000	1.0000

References

References marked with an asterisk contain useful further bibliographies

AVERINTSEV, M. B. (1970) On a method of describing complete parameter fields, *Problemy Peredaci Informatsii,* **6**, 100–109.

BARTLETT, M. S. (1964a) A note on spatial pattern, *Biometrics,* **20**, 891–892.

———— (1964b) The spectral analysis of two-dimensional point processes, *Biometrika,* **51**, 299–311.

———— (1966) *Stochastic Processes,* Camb. Univ. Press, 2nd ed.

———— (1967a) Inference and stochastic processes, *J. R. Statist. Soc.,* **A 130**, 457–477.

———— (1967b) The spectral analysis of line processes, *Proc. 5th Berkeley Symp. on Math. Stat. and Prob.,* **Vol. III**, 135–153.

———— (1968) A further note on nearest neighbour models, *J. R. Statist Soc.,* **A 131**, 579–580.

———— (1971a) Two-dimensional nearest-neighbour systems and their ecological applications, *Statistical Ecology,* **Vol. 1**, 179–194, Penn. State Univ. Press.

———— (1971b) Physical nearest-neighbour models and non-linear time-series, *J. App. Prob.,* **8**, 222–232.

———— (1972) Physical nearest-neighbour models and non-linear time-series II, *J. App. Prob.,* **9**, 76–86.

———— (1974a) The statistical analysis of spatial pattern, *Adv. App. Prob.,* **6**, 336–358.

———— (1974b) Physical nearest-neighbour models and non-linear time-series III, *J. App. Prob.,* **11**, 715–725.

BARTLETT, M. S. and BESAG, J. E. (1969) Correlation properties of some nearest-neighbour systems, *Bull. Int. Statist.,* (Book 2) **43**, 191–193.

BATCHELOR, L. D. and REED, H. S. (1918) Relation of the variability of fruit trees to the accuracy of field trials, *J. Agric. Res.,* **12**, 245–283.

BESAG, J. E. (1972a) On the correlation structure of some two-dimensional stationary processes, *Biometrika,* **59**, 43–48.

—— (1972b) Nearest-neighbour systems and the auto-logistic model for binary data, *J. R. Statist. Soc.*, B **34**, 75–83.

—— (1972c) Nearest-neighbour systems: a lemma with application to Bartlett's global solutions, *J. App. Prob.*, **9**, 418–421.

—— (1974a) Spatial interaction and the statistical analysis of lattice systems, *J. R. Statist. Soc.*, B **36**, 192–236.

—— (1974b) On spatial-temporal models and Markov fields, *Trans. European Meeting of Statisticians*, Prague.

BESAG, J. E. and GLEAVES, J. T. (1974) On the detection of spatial pattern in plant communities, *Bull. Int. Statist. Inst.*, **45**, Vol. I, 153–158.

BESAG, J. E. and MORAN, P. A. P. (1975) The estimation and testing of spatial correlation in Gaussian lattice processes. To be published.

BLACKITH, R. E. (1958) Nearest-neighbour distance measurements for the estimation of animal populations, *Ecology*, **39**, 147–150.

BROOK, D. (1964) On the distinction between the conditional probability and the joint probability approaches in the specification of nearest-neighbour systems, *Biometrika*, **51**, 481–483.

CLARKE, P. J. and EVANS, F. C. (1954) Distance to nearest-neighbour as a measure of spatial patterns in biological populations, *Ecology*, **35**, 445–453.

CLIFF, A. D. and ORD, J. K. (1973) *Spatial Autocorrelation*, Pion, London.

—— (1975) Model building and the analysis of spatial pattern in human geography, *J. R. Statist. Soc. B*, **37**, 297–348.

CLIFFORD, P. and SUDBURY, A. W. (1973) A model for spatial conflict, *Biometrika*, **60**, 581–588.

COOLEY, J. W. and TUKEY, J. W. (1965) An algorithm for the machine calculation of complex Fourier series, *Math. of comput.* **19**, 297–301.

COX, D. R. and LEWIS, A. W. (1966) *The Statistical Analysis of Series of Events*, Chapman and Hall, London.

COX, D. R. and MILLER, H. D. (1965) *The Theory of Stochastic Processes*, Chapman and Hall, London.

CRISP, D. J. (1961) Territorial behaviour in barnacle settlement, *J. Exp. Biol.*, **38**, 429–446.

*DIGGLE, P. (1973) Contagion and allied processes (M. Sc. Thesis, Univ. of Oxford).

DOBRUSHIN, R. L. (1968a) The description of a random field by means of its conditional probabilities, *Th. Prob. and Appl.*, **13**, 197–224.

—— (1968b) Gibbsian random fields for lattice systems with pairwise interactions, *Funktional'nyi Analiz*, **2**, 292–301.

—— (1968c) The problem of uniqueness of a Gibbsian random field and the problem of phase transitions, *Funktional'nyi Analiz*, **2**, 302–312.

EBERHARDT, L. L. (1967) Some developments in 'distance sampling', *Biometrics,* **23**, 207–216.

FREEMAN, G. H. (1953) Spread of diseases in a rectangular plantation with vacancies, *Biometrika,* **40**, 287–305.

FRENCH, A. S. and HOLDEN, A. V. (1971) Alias-free sampling of neuronal spike trains, *Kybernetik,* **8**, 165–171.

GHENT, A. W. (1963) Studies of regeneration of forest stands devastated by the Spruce Budworm, *For. Sci.,* **9**, 295–310.

GHENT, A. W. and GRINSTEAD, B. (1965) A new method of assessing contagion, applied to a distribution of red-ear sunfish, *Trans. Amer. Fisheries Soc.,* **94**, 135–142.

GLASS, L. and TOBLER, W. R. (1971) Uniform distribution of objects in a homogeneous field, cities on a plain, *Nature,* **233**, 67–68.

*GREIG-SMITH, P. (1957) *Quantitative Plant Ecology,* Butterworths, London.

GRIMMETT, G. R. (1973) A theorem about random fields, *Bull. Lond. Math. Soc.,* **5**, 81–84.

HAMMERSLEY, J. (1972) Stochastic models for the distribution of particles in space, *Adv. App. Prob., (Spec. Suppl.)*

HANNAN, E. J. (1955) Exact tests for serial correlation, *Biometrika,* **42**, 133–142.

HEINE, V. (1955) Models for two-dimensional stationary stochastic processes, *Biometrika,* **42**, 170–178.

*HOLGATE, P. (1972) The use of distance methods for the analysis of spatial distribution of points, *Stochastic Point Processes,* Wiley, New York (122–135).

*KENDALL, M. G. and MORAN, P. A. P. (1963) *Geometrical Probability,* Griffin, London.

KERSHAW, K. A. (1957) The use of cover and frequency in the detection of pattern in plant communities, *Ecology,* **38**, 291–299.

LÉVY, P. (1948) Chaînes doubles de Markoff et fonctions aléotoires de deux variables, *C. R. Acad. Sci. Paris,* **226**, 53–55.

MATÉRN, B. (1960) Spatial variation, *Medd. Skogsforskn Inst.,* **49**, no. 5.

——— (1971) Doubly stochastic Poisson processes in the plane. *Statistical Ecology,* **Vol. I,** Penn. State Univ. Press (195–213).

MEAD, R. (1966) A relationship between individual plant spacing and yield, *Ann. Bot.,* **30**, 301–309.

——— (1967) A mathematical model for the estimation of interplant competition, *Biometrics,* **23**, 189–205.

——— (1968) Measurement of competition between individual plants in a population, *J. Ecol.,* **56**, 35–45.

——— (1971) Models for interplant competition in irregularly spaced populations, *Statistical Ecology,* **Vol. 2,** Penn. State Univ. Press (13–30).

────── (1974) A test for spatial pattern at several scales using data from a grid of contiguous quadrats, *Biometrics*, **30**, 295–307.

MERCER, W. B. and HALL, A. D. (1911) The experimental error of field trials, *J. Agric. Sci.*, **4**, 107–132.

*MORAN, P. A. P. (1966) A note on recent research in geometrical probability, *J. App. Prob.*, **3**, 453–463.

────── (1973) A Gaussian Markovian process on a square lattice, *J. App. Prob.*, **10**, 54–62.

NEWELL, G. F. and MONTROLL, E. W. (1953) On the theory of the Ising model of ferromagnetism, *Rev. Mod. Phys.*, **25**, 353–389.

NUMATA, M. (1961) Forest vegetation in the vicinity of Choshi–coastal flora and vegetation at Choshi. Chiba Prefecture IV. *Bull. Choshi Marine Laboratory*, Chiba University, No. 3, 28–48.

OGAWARA, M. (1951) A note on the test of serial correlation coefficients *Ann. Math. Statist.*, **22**, 115–118.

ONSAGER, L. (1944) Crystal statistics I. A two-dimensional model with an order-disorder transition, *Phys. Rev.*, **65**, 117–149.

ORD, K. (1972) Estimation methods for models of spatial interaction, (presented at ASA conference, Montreal).

PIELOU, E. C. (1959) The use of point-to-plant distances in the study of the pattern of plant populations, *J. Ecol.*, **47**, 607–613.

────── (1964) The spatial pattern of two-phase patchworks of vegetation, *Biometrics*, **20**, 156–167.

*────── (1969) *An Introduction to Mathematical Ecology*, Wiley, New York.

PRESTON, C. J. (1973) Generalized Gibbs states and Markov random fields, *Adv. App. Prob.*, **5**, 242–261.

ROSANOV, YU. A. (1967) On the Gaussian homogeneous fields with given conditional distributions, *Th. Prob. and Appl.*, **12**, 381–391.

SKELLAM, J. G. (1952) Studies in statistical ecology, (I) Spatial pattern, *Biometrika*, **39**, 346–362.

SOLOMON, H. and WANG, P. C. C. (1972) Non-homogeneous Poisson fields of random lines with applications to traffic flow, *Proc. 6th Berkeley Symp. on Math. Statist. and Prob.*, Vol. 3, 383–400.

SPITZER, F. (1971) Markov random fields and Gibbs ensembles, *Am. Math. Monthly*, **78**, 142–154.

SWITZER, P. (1965) A random set process in the plane with a Markovian property, *Ann. Math. Statist.*, **36**, 1859–1863.

────── (1971) Mapping a geographically correlated environment, *Statistical Ecology*, Vol. 1, 235–269 Penn. State Univ. Press.

THOMPSON, H. R. (1955) Spatial point processes, with application to ecology, *Biometrika*, **42**, 102–115.

WHITTLE, P. (1954) On stationary processes in the plane, *Biometrika*, **41**, 434–449.

——— (1962) Topographic correlation, power-law covariance functions and diffusion, *Biometrika,* **49**, 305–314.

WONG, E. (1969) Homogeneous Gauss-Markov random fields, *Ann. Math. Statist.,* **40**, 1625–34.

YANG, C. N. (1952) The spontaneous magnetization of a two-dimensional Ising model, *Phys. Rev.,* **85**, 809–816.

Index